T0141123

Gut Anthro

Gut Anthro

An Experiment in
Thinking with Microbes

Amber Benezra

UNIVERSITY OF MINNESOTA PRESS
MINNEAPOLIS
LONDON

All photographs were taken by the author and with the express permission of the subjects.

Copyright 2023 by the Regents of the University of Minnesota

All rights reserved. No part of this publication may be reproduced, stored in a retrieval system, or transmitted, in any form or by any means, electronic, mechanical, photocopying, recording, or otherwise, without the prior written permission of the publisher.

Published by the University of Minnesota Press
111 Third Avenue South, Suite 290
Minneapolis, MN 55401–2520
http://www.upress.umn.edu

ISBN 978-1-5179-0129-5 (hc)
ISBN 978-1-5179-0130-1 (pb)

A Cataloging-in-Publication record for this book is available from the Library of Congress.

The University of Minnesota is an equal-opportunity educator and employer.

UMP BmB 2023

For Lydia, Beatrice, and Joe

Contents

Preface

I had a miscarriage on an airplane. I was returning home from a month of fieldwork in Bangladesh, having spent eight hours a day, seven days a week interviewing Bengali mothers who were fighting malnutrition with every emotional and material tool they had. Only my translator, Israt, and a couple of the other women field research assistants knew I was pregnant. When I left Dhaka, they gave me a red clay sculpture of a mother and a baby—even though in the neighborhoods of Mirpur they know about the fragility of hope. The women I knew there didn't talk publicly about their own pregnancies; there are superstitions about giving gifts before the baby is born. Bad luck.

A few hours into my second flight, from Dubai to New York, I began to bleed. I bled more than I ever have in my life, was in more pain. I panicked and cried and refused to believe I was losing my baby, even as I lay on the floor of the airplane bathroom in the fifth, eighth, tenth hours of a thirteen-hour flight. No one tells you that a miscarriage can be like giving birth, that you have contractions, that your uterus pushes out the fetus, even if it is only ten weeks old. No one tells you that you will keep bleeding and bleeding for hours and that huge pieces of tissue, of placenta and solid matter, will also come out of you. No one tells you how much it hurts your body. I deliriously searched the airplane toilet for signs of a baby. Reached in with my hands and pushed my fingers through the soft mess. How big is a baby at ten weeks old? One inch? The size of a kumquat? Too small for me to find in the dirty blue bowl in the economy class toilet of Emirates Airline flight 205. The flight attendants asked over the intercom if there was a doctor on board. They found me a seat in business class where I could lie down; they were sympathetic to my partner and me, as was the doctor, who kindly lied and said she was sure everything was going to be OK. The ensuing events were a blur—taken by ambulance from JFK to a hospital in Jamaica, Queens; freezing in a hospital gown for eight hours; a wordless ultrasound; the disastrous insertion of an IV that sent a spray of blood from my arm to the stained ER ceiling; peanut M&Ms from a vending machine that would be the only food I ate for two days; a cruelly dispassionate question from an ER doctor: was I sure I had been pregnant? Because there was definitely no baby in my uterus now.

I write about this because this miscarriage made it hard for me to write this book until now. I was bereft, and the experience irrevocably impacted my relationship to my fieldwork. My project is largely about mothers and babies. Yes, it is about microbiome science and the ways in which microbial life has come to figure importantly in anthropology as an ontological link between the biological and social sciences. But many of the most important actors in this story are mothers and babies. And this unbelievably painful event became tangled up in my fieldwork and put my own flesh into the story occupied by so many other maternal bodies. It changed what is important to me and what I cared about in this project. It changed the type of engagement I had with the Gordon Lab and the ways I approached the scientific work. It's important to say I was fieldworking collaboratively with my spouse, Joe DeStefano, who was also an anthropologist in the employ of the Gordon Lab. Joe was with me in Dhaka, and we performed the lab ethnographic work together. The very complicated personal and professional entanglements crisscross not only Jeff Gordon and me but Jeff and Joe, and Joe and me.

It turns out that having children unexpectedly unravels our carefully wound theoretical commitments. After the miscarriage and during later pregnancies and births, I watched myself run full speed toward biomedicine for answers, certainty, and help, unnerved and unsettled, in my mind temporarily renouncing all the intellectual work I had done *against* biomedical authority. Emily Yates-Doerr has written brilliantly about what she wanted when her child was injured during her own fieldwork—the history of mothers' knowledge, but also X-rays, anesthesia, and surgery:

> I have used the conjunction "and" here, because my child's fall and the support I was surrounded by afterward helped me understand how wanting care and wanting technology are not inherent contradictions. They can go together. Mothering and doctoring. To be clear: I am not advocating for magic bullet fixes to problems asking for structural-level responses . . . nor am I leaving biomedicine unchallenged. I am, rather, engaging with science and technology's multitudes.[1]

Similarly, Jody A. Roberts does a smart and moving analysis of his daughter's cerebral palsy and how to live realistically with critical science and technology studies (STS) perspectives in the real world: "We've stood on the margins in order to celebrate the peripheral while criticizing the center for believing it is the center. We've drawn networks of interest, influence, and power, but failed to fully engage the fact that we, too, are parts of these networks."[2] Moreover, we desperately need these networks and multitudes when we are damaged, terrified, and hurt.

The miscarriage fostered a deep emotional bond between me and my scientific collaborator, who compassionately supported me through the psychological and physical aftermath. And it affected us later, as our professional relationship broke, confused by undercurrents of feared betrayal, personal disconnections, and hurt feelings. It made me question my decisions and my privilege—to travel while pregnant to a country with extraordinarily high levels of air, water, and soil pollution, a place where food deemed in the United States as "prenatally appropriate" was nonexistent. To travel to that place, and then to voluntarily leave it. I was wracked with guilt that I had shirked my biomaternal responsibility in exchange for my fieldwork, that I endangered (and ended) my baby's life to further my professional career. I obsessed about the microbes I had exposed myself and my baby to in traveling across the world and working in unsafe housing, water, and sewage conditions. And subsequently, I questioned why my uterus should be more valuable than the uteruses of all the mothers with whom I worked in Mirpur, women to whom those toxins, pathogens, and precariousness are a condition of everyday life.

It is risky business, talking about my miscarriage as a woman academic. Doubly risky, as first, I may invalidate my own scholarly status with sentimentalism, and second, I risk myself by laying bare my experience for academic consumption. There have been anthropologists who have taken miscarriage as a social process and their ethnographic objects.[3] There are some before me who have beautifully and boldly discussed their own bodies as crucial in framing their anthropological analysis, their own biological entanglements with their research.[4] It was risky for them too, the peril of appearing unprofessional, emotional, too subjective. But I can't leave it out of the story; this terrible thing happened to me in the field and is part of my fieldwork. In describing feminist health research in the context of her own reproductive experiences, Rayna Rapp has said, "Our stories are unevenly and globally scripted. . . . Learning to live with the benefits and burdens of multilayered nondisclosure has thus been far from academic."[5] The children I eventually had are also part of the story—these daughters, my own biology, and all the mother/child/woman involutions that rerouted my professional career, both because of the personal choices I would make and because of the powerful patriarchy that persists in academic anthropology. The future of ethnography is one of partially connected bodies, where kinship is not only the object of study but a process that we live within.[6] The ethical commitments of feminist anthropology, the particular questions we ask of our ethnographic objects, our collaborators, our field sites, and ourselves, gnaw at and compel me. Anthropologists must talk about the familial, intellectual, and bodily violences we negotiate as women and gender-diverse fieldworkers. And even as we talk more about sexual harassment and gender discrimination that happens during fieldwork,[7] we remain silent about those same violations during campus visits, conference panels, advising meetings, and colleague dinners.[8] We are obligated to unfalteringly question the white supremacy, heteronormativity, ableism, and sexism of academia—especially in places like sociocultural anthropology, where harm hides behind disciplinary self-righteousness. Somehow still, the majority of anthropological work insists on resisting a feminist anthropology, marginalizing an equitable race/gender/ability framework. In questioning critical turns toward theories of contamination and away from contaminated communities, David Bond wonders about "who we write for and why."[9] Who *do* we write for? We must examine the representational violence we continue to participate in to make careers for ourselves,[10] our convenient disconnections from the very real violence within which our interlocutors sometimes live. As a science-collaborating medical anthropologist, I operate within a never-ending negotiation—always attuned to the ways science handily exchanges life and death circumstances for data and the ways in which anthropology does the same for theory and clever neologisms.

In my microbiome work, Bangladeshi mothers bear the weight of international translational research. They provide the fleshy materiality for the science; they give from a wide geography of their own bodies and the bodies of their babies: blood, milk, feces, and microbes. They live their own complicated lives while also making the science (and ethnography) possible. They don't get paid or get credit; they care primarily about keeping their kids alive. During this research, parts of my own viscera, pulped and destroyed, were muddled into this work. As Carrie Friese has described: "as my research has bled into real life, I have found my real-life bleeding back into research."[11] For me, the blood was literal. Everything was connected—my miscarriage, my project, the microbiome study, myself as a mother fieldworker. I did not write this book under the auspices of a writing fellowship or sabbatical, luxuriating in protected time and space. I wrote it after bedtime and on weekends, in the other snips and bits of time in between breastfeeding and teething, croup, puke, and shit. I wrote while I was keeping young and old humans alive—alongside my children, I cared for an aging parent with Alzheimer's and dementia. *Care* is too tender a word here; more accurately, I drowned under my responsibility to other lives. I neglected to care for my own body as some of my parts grew monstrous and diseased, all the time teaching, publishing papers, and grinding on the job market. Many of these were intentional choices I made and would not take back, and I recognize my race, class, and ability privilege in a system that made all of this (hard but not im)possible. I write about these conditions because they are the labor that academia renders invisible,[12] things we are expected not to mention, that we are expected to succeed in spite of. In fact, these tribulations were the scaffolding for the anthropological work. Perhaps as Julie Johnson Searcy and Angela N. Castañeda write, for me mothering itself became a kind of collaborative expertise that transformed the academic process.[13] I was not intellectually impartial about my ethnographic objects or interlocutors; I became attached to the people with whom I worked, I lost a baby, had two other babies, lost two fathers, and was devastated by the eventual dismantling of my anthro/science collaboration. This very book was paralyzed into stasis for years by various griefs. This anguish drove me to the imperative I feel now, to *do anthropology differently.*

If we are to enact the necessary revolution of the social and biological sciences, one that operationalizes transdisciplinary collaborations in ways that meaningfully address suffering on this planet, we must question and dismantle our own rigid allegiances. These dangerous investments can be to any number of toxic binaries: race, gender, sexuality, nature/culture. The tenacious fealties this book tries to tackle are those committed to maintaining divisions between categories of human and nonhuman and those that hold anthropology as ethically superior to biological science. These are mutually interdependent ideologies—as long as you can neatly categorize "people," you can assert that anthropology cares more about them than biology does. *Gut Anthro* hopes to unsettle those divisions, using microbes as a vector of disruption.

The grandmother of one of the study children gets ready to prepare *shaak* (leafy greens) for her family.

Introduction

I have seen a lot of guts, and my fair share of shit. I have been studying human microbiomes and the scientists who study microbiomes for over a decade. Human microbiota is the community of microbes in and on us, and the human microbiome is the collection of microbial and human genes in a dynamic and interactive microecosystem that changes over time.[1] In our guts, we have the largest, most diverse, and most powerful collection of microbes, many more microbial cells and genes than human ones.[2]

I worked collaboratively for several years with the Gordon Lab at Washington University, a preeminent human microbiome research laboratory developing translational pipelines between microbes and malnutrition. My ethnographic fieldwork was in the lab and at the scientific research site in Dhaka, where a Microbiome Discovery Project was being conducted in a birth cohort of hundreds of mothers and children. My job for the lab was to contribute qualitative data about Bangladesh and Bangladeshis to the microbiome science. Many of the Gordon Lab experiments use the gut microbes of malnourished babies in Dhaka, trying to figure out connections between bacteria, food, and illness. I have seen guts when the in becomes the out: germ-free mice sacrificed, blood and epithelium exposed to air, tiny stomachs split and scraped. I have seen people with guts in pain: a confluence of poverty and diarrheal disease so severe and continuous that a body can't stay nourished. I have seen what guts produce: bacterial metagenomes sequenced from human feces to map a disease and nutrition landscape. With mice, the scientists can take a gut apart and track different microbial ecosystems, but as a postdoc at the Gordon Lab told me, "in human samples, feces are all you have. You have to make a story from that."

Vanessa Ridaura is pulverizing frozen human shit. She works the glass pestle quickly against the side of the container, breaking large brown chunks into smaller ones. "If the samples thaw, it won't degrade the DNA, but it will be gross," she says, working as carefully and quickly as she can. I help her sort and organize the tubes. Bleach, UV light, or soap will kill bacteria and destroy their DNA, and so the shit can't be purified or cleaned. The samples are still several steps away from being molecularized or metagenomically analyzed, so at this point, the shit must remain shit-like. The properties of the samples, especially the color and smell, make it hard to forget what it is. Vanessa was a PhD student in the Gordon Lab at the Center for Genome Sciences & Systems Biology at Washington University. Scientists in this lab work extensively with germ-free mice: mouse bodies, blood, tissue, and lots of mouse shit. But human shit is different. Vanessa knows that by mashing it, she is homogenizing the sample—she's fairly certain there are different bacteria in different parts of the gut, and probably even different parts of the shit. But for this experiment, she just needs to get it broken down without getting too grossed out.

Shit tells a story about what microbes are living in the human gut and what they're doing there. Jeff Gordon, known as the father of the microbiome,[3] and the Gordon Lab have introduced the idea of the gut as a metabolic organ, with tens of trillions of bacteria playing a crucial role in defining humans' nutritional status, immune function, and overall illness or health. These scientists want to understand the foundations of human relationships with these microbes; how they are affected by diet, genes, and environment; and whether or not the microbial communities can be intentionally and durably altered. The shit is a means to this end—for me too. As an ethnographer, I tracked the origins of the shit, from where and from whom; what the microbes meant to people; and how microbes were ontologically enacted, through cutting-edge science, sequencing and bioinformatic technologies, global public health programs, and probiotic interventions. In collaboration with scientists, I'm trying to hold microbiome science accountable to the sociomaterial, political, and economic conditions of life while working out tactics for integrating ethnographic information into the design and implementation of human microbiota studies.

Microbes populate our bodies by the hundreds of trillions, from the moment of birth until long after we've died. Biologically, materially, and genomically, we are much more microbe than human. There are more microbes in one person's mouth than there ever were human beings on earth, more microbial organisms on earth than stars in the universe. They are on every continent, in every ecosystem: in the air, soil, and water, in the deepest oceans and hottest volcanic vents. Human–microbial configurations are constitutive of our life on earth. Human microbiomes inextricably entangle biological processes with social, intimate human practices, and consequently, the study of human microbial ecology also entangles the life and social sciences. Microbes are a vector of coevolution across life-forms and between disciplines.

In this book, I develop an *anthropology of microbes.* An anthropology of microbes mines the collaborative capacities of ethnography and feminist science studies to actively contribute to scientific projects—partnering with scientists to develop more holistic analytical and methodological frameworks to understand the embodiments of biology within social dynamics. I ask, what would it mean for anthropology to *act with* science? An anthropology of microbes operationalizes ethnographic knowledge, asking us to be ethnographers *of* and *for* scientific research, facing the corresponding compromises and uncertainties, challenges and failures. Feminist anthropology has always been at the forefront of scrutinizing and unpacking how scientific knowledge affects bodies, makes gender, and produces difference. The future of the close and contentious relationship between feminist anthropology and microbiome science has the potential to question what is meant by nature, culture, the social, and the biological, examining the ways in which microbial life has come to figure importantly in anthropology as an ontological link between the biological and social sciences.[4] This work also mines shit—both literal feces and unpleasant collaborative complications—for scientific and ethnographic gold.

Rather than mere observational fieldwork among scientists, an anthropology of microbes requires anthropologists to become members of scientific teams and contribute ethnographic data, while requiring biological scientists to make space in research design, experimentation, and implementation for social science expertise and thinking. These transdisciplinary collaborations are not easy; the feminist anthropological stakes are continuously offset and commingled with concrete questions of health, illness, and biology, as well as highly technical data. Asymmetries of power, anger, and frustration abound. An anthropology of microbes is the product of a circuit of collaboration that crosses class, racial, and national boundaries—and, for me, unfocuses the lines of individual, personal, and professional agency. I endeavor to reframe sociocultural anthropology's relationship with (big *S*) Science, scientists, and microbes—a reformation that has already started to take place between biology and science studies, public health, geography, and sociology. This is an attempt to move transdisciplinary engagement beyond ethical policing, theoretical backbiting, and critical observation toward action—practices in the world predicated on an agreement that, "at all levels, the biological and the social are in one another."[5] The theoretical scaffolding of feminist theory and ethnographic data of anthropology may be able to provide a more gradated context through which to enact ethical and effective life science, while social scientists learn to take material biology and the corresponding science seriously, compelled by practices, outcomes, and action. An anthropology of microbes is a coevolution, transforming human microbial ecology *and* anthropology through the operation of uneasy partnerships between humans and microbes, social and biological scientists.

The "Anthropocene" finds us immersed in a collection of crises: human beings on the brink (or at least imminently aware) of simultaneous climate/environment, food/nutrition, race/gender, and violence/inequality cataclysms. Whether or not the Anthropocene is a valid state of being is vigorously contested. To trust in the Anthropocene is to acknowledge the deadly, destructive impact humans have had on the air, land, water, and organisms of earth, while at the same time wholeheartedly reinvesting in human exceptionalism. It has become what Elizabeth Reddy calls a "charismatic mega-category"[6] that anthropologists seem eager to embrace, conglomerating conflicting discourses without asking, whose apocalypse is it anyhow? Zoe Todd problematizes the center of the Anthropocene: "What does it mean to have a reciprocal discourse on catastrophic end times and apocalyptic environmental change in a place where, over the last five hundred years, Indigenous peoples faced (and face) the end of worlds with the violent incursion of colonial ideologies and actions?"[7] And forcing us to think about legacies of extractive imperialism and racism, Kathryn Yusoff says, "If the Anthropocene proclaims a sudden concern with the exposures of environmental harm to white liberal communities, it does so in the wake of histories in which these harms have been knowingly exported to black and brown communities under the rubric of civilization, progress, modernization, and capitalism."[8] It is time to ask, who gets to be human, and how does that category matter? Threats of extinction are stratified. Indisputably, a different kind of shit is hitting the fan.

Transdisciplinary cooperation between the social and biological sciences has never been more important and, at the same time, never seemed more impossible. Since 2009, I have been working on the human microbiome in collaboration and conflict, love and hate, sickness and health, with human microbial ecologists. Anthropological colleagues, peer reviewers, and senior faculty have called me "a handmaiden of science," accused me of "losing my critical edge," and written me off as having "drunk the lab Kool-Aid" (all actual quotes, addressed to me). Scientific allies have fretted about misinterpretations, conceptual errors, and confusing combinations of jargon, while demanding procedures to formally sign off on anthropological work containing personal experience and analysis alongside scientific concepts. I failed on both sides—didn't fit in, didn't get published, didn't get jobs, didn't get funded. I cautiously sought a home in STS.

The prefix *inter-* accompanies and troubles me everywhere I go in this work: *interspecies, interacting, intergenerational, interlocutors, interfere, intervention, interdependence, interrelationships, interconnected,* and, most of all, *interdisciplinary.* Semantically, *inter-* means between or among, mutually or reciprocally. In my anthropology of microbes, *inter-* means all that and also "with, and, in the middle of," sometimes painfully.

In important ways, this book is about the process of collaboration, what it means to collaborate—with Bangladeshi mothers, local field researchers, massive philanthropic global health organizations, the Gordon Lab scientists, and, of course, microbes. But biosocial collaboration requires one to get fully face down into the shit. There is no such thing as easy, effortless interdisciplinary work. My anthropological analysis always came from a "soft" science, using "weak" language, but a perspective that was also sometimes perceived to be a threat—stripping clothes from the emperor, exposing the flawed master narrative of scientific and ethical certainty. My fieldwork in Dhaka and St. Louis shows how work across disciplines (including anthropology) produces the microbiome as an experimental object and provides descriptions of how ethnography and microbial ecology interfered with, enriched, or otherwise changed each other. The goal of this work was a partnership between anthropology and microbial ecology that tried to lead to a deeper understanding of the microbiome and better solutions for addressing malnutrition and enteric disease. Changes to the ways humans live on earth and interact with environments and corresponding exposures, even the routes by which we are born and which foods we put down our collective gullets, are affecting what collections of microbes we have and don't have in our guts. Human microbial ecologists studying shit worry that humans are sliding toward a singular, globalized, high-sugar, antibiotic-ridden lifestyle, and our previously diverse population of microbes is sliding with us. Science's worry is somehow both without history and in denial of their own complicity. In analyzing the self-made scientific futures of antibiotic resistance, Hannah Landecker comments, "Bacterial life today is appearing as a specific instantiation of the biology of the Anthropocene: human efforts to control life's productivity become the matter of the world."[9] Anthropocenic damage is not just macro; it is also micro.

Accounting for microbes troubles the waters of inside–outside, biological–social, community–individual. Or perhaps these waters were already disturbed and white academia is characteristically late to the game. Because anthropology and non-/post-/transhumanism remain mostly white social and intellectual spaces[10] that reproduce racism in theories, research, and departments,[11] who "we" are isn't clear, and who gets designated as human or nonhuman isn't a given. Moves are being made toward the decolonization of anthropology, ontology, and studies of interspeciality.[12] This work exposes the harmful intersection of environmental/racial/gender violence, cis-heteropatriarchy, and anthropocentrism and makes crucial points: the white supremacy of the academy, structures of publication and citation, and ways of producing knowledge are epistemically violent in their exclusions. Those exclusions are not just intellectual but tied to real-life violence, sovereignty, and political and bodily harm—and cursory mentions or appropriation of Indigenous thought are a continued form of colonization.[13] Post-/non-/transhuman theories have been debating the agency, sociality, and ontologies of microbes and things like microbes, all the while appropriating and eliding Indigenous scholarship that directly addresses the nonhuman world. I hope this book can begin to evoke Indigenous formulations that necessitate reciprocal, ethical accountability to more-than-human relations.

I became interested in microbes because I was excited about how they translated, how they mapped onto anthropological ways of asking questions of the world. I wasn't prepared for the disciplinary shit I would become mired in—the ways in which microbial considerations forced changes in the ways I thought about and practiced anthropology. Studies of microbes necessitate a micro-view reorientation to concepts of ecology, geography, nutrition, and data, a critical proximity to the details of micro-objects as well as attention to the macro-processes within which they are implicated. This book uses autoethnography to illustrate the set of conflicts, experiments in methodological hybridization, boundary frictions, and misfires of an anthropology of microbes *in process.* As the Gordon Lab scientists searched through Bangladeshi shit for answers to microbiome mysteries, I maneuvered through a collaborative labyrinth of shit, through twists and turns I couldn't have anticipated and had no map to navigate.

Origin Stories

In spring 2009, I wrote a handful of emails to the principal investigators (PIs) of the newly National Institutes of Health (NIH)-funded Human Microbiome Project (HMP). The HMP was established to generate resources, technologies, and programs to comprehensively characterize the human microbiome and analyze its role in human health and disease. Subsequently, having summarily failed in characterizing a "normal microbiome," the HMP began its second iteration in 2012. As of 2014, the iHMP (Integrated Human Microbiome Project) was tasked with creating integrated longitudinal data sets from both the microbiome and host from three different cohort studies of microbiome-associated conditions using multiple "'omics technologies."[14] I had circuitously become interested in the HMP as a PhD candidate developing a dissertation project about new genetic technologies and kinship and the ways in which genomic testing was transforming ideas of human relatedness. Most of the HMP PIs didn't reply to me at all, but surprisingly, Dr. Gordon responded with a phone call.

At the time, I had no idea how extremely busy his life as an internationally known PI was. I didn't know Dr. Gordon at all, or that he would exert extraordinary effort to make and maintain intensely personal-professional relationships. In that first conversation, Dr. Gordon and I talked for an hour. Unlike the unreceptive, uncaring scientist I expected, he was open, kind, and enthusiastic. He was genuinely concerned with how humans did science and what science did to humans. He said, "I cannot stress more emphatically how this work needs cultural anthropologists." He closed our conversation with an invitation to come to St. Louis and visit the Gordon Lab. Instead of showing resistance and doubt, this scientist wanted to fund a site visit and was eager to have me explore his work. Was my graduate school training (and learned suspicion and distrust) based on an outmoded formulation of the real-world relationships between biological and social scientists? Yes, and no. The next seven years would be an astounding, infuriating, illuminating, and exhausting experiment in building a relationship with this singular scientist and his lab and in figuring out the terms of our collaboration. The sea of shit kept rising; I found myself knee-deep in confounding situations that were at times intimate *and* professional, scholarly, financial, and political.

When I first met Dr. Gordon in August 2009, his lab was at the Center for Genome Sciences & Systems Biology at Washington University School of Medicine, on the fifth floor of an unassuming office building on Forest Park Avenue, named for the thirteen-hundred-acre, 137-year-old public park a few blocks away. The building had hotel-style industrial carpeting throughout and potted plants on every floor. Dr. Gordon's office was casual and comfortable, not typically academic—no overflowing bookshelves, no piles of papers. He had a large fish tank; an Apple cinema display; a digital picture frame with changing photos of animals; awards, degrees, and accolades; a framed Cathy cartoon; and a metal plaque that said "*ancora imparo*" (I am still learning). The Gordon Lab's modest floor contained an autoclave room, a metabolomics facility, anaerobic chambers, a robotics room, −80°C freezers (filled to bursting with samples of shit), PCR machines, a human sample coy chamber, an ultracentrifuge, a room for making mouse food, a tissue culture room, a radioactive room, and a phage room. In 2016, the Gordon Lab moved into a new $81 million research facility, designed for Leadership in Energy and Environmental Design (LEED) silver certification, and state-of-the-art, highly flexible open labs. It is now known as the Edison Family Center for Genome Sciences & Systems Biology, and Dr. Gordon still serves as the director.

The Center for Genome Sciences & Systems Biology at Washington University School of Medicine, 2009.

In 2010, the Gordon Lab, in collaboration with the International Centre for Diarrhoeal Disease, Bangladesh (icddr,b), was studying the interrelationships between diet and microbial community structure/function and whether differences in human gut microbial ecology affect predisposition to malnutrition in children. An international health research organization as well as a free public hospital in Dhaka that serves hundreds of thousands of patients yearly, icddr,b is known locally as the "cholera hospital" and is the world's largest diarrheal disease hospital. Founded in the early 1960s, icddr,b administers the world's longest-running health and demographic surveillance program, a dominating research presence in Bangladesh with complicated histories and politics of care. The organization has a colossal, sometimes prevailing presence in the communities it serves (more about this in chapter 2). Bangladesh's subtropical monsoon climate has three distinct seasons with great variation in rainfall, temperature, and humidity. Summer is March to June and hot and humid, followed by the extremely rainy and cool monsoon season, and October to March is winter, dry and cool. During the monsoon, or "cholera season," icddr,b serves hundreds of people a day, far exceeding its 350-bed capacity. From June to October, patients spill out into the temporary cholera triage unit in a loosely tented area in hospital parking lot. Triage in the eye of a cholera storm means rows and rows of plastic cots with holes in the beds, buckets underneath to catch the uncontrollable amount of shit. In Dhaka, shit is a menacing force; it violently overwhelms bodies, families, and communities. Having power over shit can mean the difference between life and death.

Funded primarily by the government of Bangladesh, icddr,b receives money from five core donors[15]—the Australian Agency for International Development (AusAid), the United Kingdom's Department for International Development (DFID), the Netherlands, the Swedish International Development Cooperation Agency, and the Canadian International Development Agency (CIDA)—as well as fifty other nations and nongovernmental organizations (NGOs). The Gordon Lab and icddr,b are partners in the global Mal-ED consortium. Pronounced "mal-a-dee," Mal-ED stands for Malnutrition and Enteric Diseases, a foundation of the NIH-run, Bill and Melinda Gates Foundation–funded group of U.S. research institutions and international health organizations. Mal-ED is investigating how intestinal infections alter the gut's ability to absorb nutrients and the interrelationship between malnutrition and enteric disease among children. International Mal-ED sites include Brazil, Peru, Nepal, South Africa, Tanzania, India, Pakistan, and Bangladesh. The Gates Foundation has contributed approximately $30 million to Mal-ED, and it promotes the very specific goal to "eliminate the gap in mortality from enteric and diarrheal diseases between developed and developing countries and to significantly reduce impaired development associated with these diseases in children under age five."[16] The Gates Foundation was the primary funder for two phases of the Gordon Lab microbiome–malnutrition work during my collaboration with the lab.

The Mirpur (a *thana,* or subdistrict, of urban Dhaka) field site in Dhaka is five square kilometers and houses fifty thousand residents, many of whom are rickshaw pullers, garment workers, and mothers. To these Bangladeshis, gut microbes are experienced

The main hospital of icddr,b is called the Center. The Nutritional Rehabilitation Ward is located here, and this is where all the study samples are processed. Only urgently sick, severely malnourished kids will ever come to the Center; it's likely none of the Mirpur families will ever make the ten-kilometer journey. More affordable, human-powered rickshaws aren't allowed on the highways, and the traffic is so intense that the trip takes hours on a bus or in a CNG (a motorized compressed natural gas vehicle).

as chronic diarrhea, small babies, and ubiquitous NGO public health intervention. As an ethnographer, sometimes following microbes was the physical act of going from place to place, accompanying actual bacterial cells from country to country. It meant tracing a network, attending presentations where Gordon Lab scientists attempted to explain how mouse diet experiments and metagenomics translated into ambitious solutions for global malnutrition. Other times, it meant crouching in the searing Dhaka sun next to a Bengali mother cutting fish, considering the invisible universe of microbes in the fetid open drains, and, later, sitting in over-air-conditioned conference rooms for weekly lab meetings to see those "same" microbes arriving on a PCOA[17] plot. This book explores what ethical and material practices surface at different moments in space and time within this tenuous network. As microbes move from place to place, they change—sometimes those changes matter, and sometimes they don't.

An anthropologist had never been invited to participate so fully in a major transnational microbiome study. Moreover, Jeff Gordon's lab is arguably the most preeminent and influential site of human microbial ecology research, making it a terrifying, heady, and powerful place for a PhD student in anthropology to conduct her fieldwork. *Gut Anthro* tracks my participation in this intensive "experimental entanglement"[18] over the course of seven years, two continents, and countless emotional states. Through this work, I attempt to make three moves. First, and most importantly, I set out to see if a feminist, anticolonial, antiracist anthropology can act with and produce a feminist, anticolonial, antiracist microbiome science. Is it possible to ontologically elide the division between anthropology and microbial ecology, just as the nature–culture and social–biological divides have been interrogated by other anthropologists coworking with scientists?[19] I want to fight for science as an anthropologist, but I want a different kind of science to foreground equity and ethics for all those embedded in the scientific work. Second, I draw the material agency of and relations with microbes productively together with engaged theories of social suffering. How can all that social science has theorized about post-/trans-/nonhumans be practically applied to issues of structural violence? How do microbes matter, bodily, socially, and scientifically? Third, this story is largely told through my own inner conflicts, uncertainties, regrets, and misunderstandings. The end of my relationship with Dr. Gordon left me, for several years, wounded, indebted, and disempowered, afraid and unable to write about my fieldwork. This book attends to why I was hurt and why I am still hopeful.

As lab scientists studied the interrelationships between gut microbes and malnutrition, I explored ways to reconcile the scale and speed differences between the lab; the intimate biosocial practices of Bangladeshi mothers and their vulnerable children; and the looming structural violence of poverty, abysmal sanitation, and profound inequality. I never meant to do the sort of fieldwork I did. It was the almost absurdly traditional, white-anthropologist-out-of-place-in-the-Global-South-babies-and-mothers-on-the-verge-of-making-kinship-charts ethnography that I eschewed in my PhD studies, that I refused and criticized. But I was compelled by the scientific mandate. I am not an expert on Bangladesh, nor am I a South Asianist. I don't speak Bengali. But I couldn't *not* go. It was important for me and important for the lab. Someone had to go. Should it have been the lab immunologist? The microbiologists? At least I was a burgeoning social science expert, an ethnographer who was learning to exercise her ontological commitments, practicing an educated correspondence with the world.[20] Specifically, I followed commensal bacteria in the human gut as they traveled between the field (maternal and child nutritional intervention) and the lab (microbial genomics and diet experiments). Simultaneously, I served as the laboratory's social scientist, contributing ethnographic data to analyses of the microbial genomes of Bangladeshi children suffering from severe acute malnutrition. Commensal microbes are transmitted and cultivated between humans through biological as well as social, intimate practices such as feeding, affection, and birth. I began to learn how social disruptions can cause microbial perturbations, and microbial perturbations cause social disruptions, all those disturbances with roots deep in piles of shit.

A Brief Qualification

Before continuing, the predicated-on-science claims throughout this book require redress and qualification. Solid arguments have been made about microbiomes as ontological objects created by microbiologists and anthropologists alike: microbes as model ecologies, appealing for their entanglements and multiplicities; microbiomes spuriously perceived to be independent from the technologies that bring them into being; microbes made social through "evidentiary symbiosis" with scientific experimentation.[21] These critiques worry about regression to scientific determinism and ask us to carefully consider what microbiomes are before we run amok pursuing idealized theories of microbial connection. Anthropologists Heather Paxson and Stefan Helmreich worry about the focus on the materiality of microbes (especially in terms of microbial agency) and warn against "new reductionisms" and a return to a scientific determinism they see emerging from taking microbiome science for granted. Instead, they focus on the model ecosystems that microbes represent: "The microbial realm, shared across scales and contexts, variously and simultaneously universal, ubiquitous and unique, has become a fresh court of appeal for those who would model new modes of living with and within biological nature. The question is not simply 'what is life?' but rather, 'what forms of life do we wish to insist upon?'"[22] Helmreich develops these ideas further in *Sounding the Limits of Life: Essays in the Anthropology of Biology and Beyond,* constructing the figure of the microbial human, *homo microbis.* Again, taking an anti–new materialist position, Helmreich applies scientist Jonathan Eisner's description of "microbiomania" to science studies scholars: "the microbiome is a novel kind of object or figure in biology, to be sure, but its multiple meanings do not themselves follow from the fact that microbiomes are composed of a multiplicity of organisms."[23] Concerned that biology must not be seen as "speaking for itself," Helmreich redraws the same lines between biology and culture, human and nonhuman, that microbial ecology seeks to blur. Conversely, others have protested that sociality is not exclusively human, that leaving microbes out is our hubris and exceptionalism.[24] These analyses warn against reifying the biological–social boundary. In practice, what I learned in the lab and in the field is that when scientists are your interlocutors, you must take the science seriously, as well as the biological–social divisions that are real to them. That is not the same as a reductionist return. I do not seek to hierarchize humans over microbes, or vice versa. The technologies and practical futures of microbiome research may address concerns about disarticulating biology and culture and those about rematerializing matter and move toward finally developing a conceptual tool kit in which we can "take biomedical data seriously but not literally."[25] Ethnographies of human–microbe relations have strong foundations in concepts like microbiopolitics,[26] but my investments are a bit different. I ask, What do we do with the gender/race/colonial violences built into scientific knowledge and practice? How do you defend anthropological critique on the ground? Can we see the human–microbe/biological–social as co-investments? I follow geographer-scientist Max Liboiron (Métis/Michif), who describes doing research in a space between that operates inside an incommensurability of dominant research paradigms and resistance/anticolonialism.[27]

In many ways, the heart (or guts, if you will) of my anthropology of microbes is made of people: lives, communities, practices, and bodies.[28] My work is not uncritically science facing, nor am I willing to jettison what scientists are saying about our interspecies entanglements—especially because scientific studies are happening with *real* (microbial and human) bodies in *real* life, not just at conference panels and in edited volumes. Much of the most conspicuous recent social science microbiome research has focused on microbes as distinctly nonhuman and particularly on discursive practices surrounding microbes.[29] The pursuit of decentering humans has the unwitting side effect of diffusing and depoliticizing the urgency of our current crises. I have less and less interest in spending time in scholarly debates about who is right or who is more clever. As useful as theory can be, as Paul Farmer said, "the *real* fight is against poverty and injustice. And there are lots of ways to have a clear analysis of it, and to have a strategy to addressing poverty and injustice in lots of different ways. . . . And it's important to have ideas. There are [*sic*] a lot of power in ideas and concepts. But that's not the *only* thing that we should do. We should also be very concerned with the pragmatic needs of people all around us, and as I have said those are food, food security, basic health services, public safety."[30] I take these concerns seriously. Mine is productively different from other anthropological studies of microbes because of what is at stake.

Making Microbiomes

In the study of human microbial ecology, *microbiota* are collections of microorganisms that make up the communities on various human body habitats, and *microbiome* refers to the collective set of microbial genes in these communities. It is estimated that there are at least as many microbial cells in our bodies as human cells, and the number of genes represented in our microbial communities likely matches our twenty-three thousand *Homo sapiens* genes with hundreds of millions of microbial genes. More than ever, human beings are considered microbial ecosystems, comprised of bacteria, archaea, eukaryotes, and viruses with whom we have coevolved. The human microbiome contributes essential functionalities to human physiology and is considered essential for the maintenance of human health.[31] From a biological science point of view, humans are supraorganisms (a system of multiple organisms functioning as one) or holobionts (a singular ecological unit made up of symbiotic assemblages), composites of human and microbial selves. We have evolved from and with these microbes. Human microbiomes inextricably entangle biological processes with social practices; microbial populations—affected by how and where we are born, what food we eat, who we live with and love—are digesting our food, training our immune systems, interacting with our states of health and illness, moods, and behavior.

Metagenomic technologies (on which microbiome science depends) propel future research; raise bioethical and privacy issues; disarrange logics of funding and regulation; and complicate categories of species, community, and self. As human microbial ecologists seek to explain interrelationships between microbes in the gut, mouth, vagina, and skin to autism, drug efficacy, postnatal health, and cancer (to name a few), a microbial focus materializes for sociology, anthropology, philosophy, and public health concerning issues of environment, genomics, and molecular biomedicalizations. Interdisciplinary insights have the potential to translate into the application of social science tools in the design of observational and interventional microbiome studies. As Dr. Gordon, Joe DeStefano, and I have written elsewhere, "bringing anthropology and human microbial ecology into a meaningful dialogue allows for new modes of collaborative research. It should create a symbiosis that enables both fields to co-develop in ways that encourage a more profound view of our 'humanness'—transforming our categories of 'community,' 'individual,' and 'life,' and in the process helping to address major global health inequities."[32] Microbiome research creates spaces for social and biological scientists to jointly reckon with the sociomaterial conditions and "local microbiologies"[33] in which microbiomes are enacted.

Though microbes have been on earth 3.49 billion years longer than *Homo sapiens,* for humans, there is no history of microbes separable from a history of microbiology. Most reductively, microbes are microscopic organisms; that is, they require a microscope to be seen by the human eye. Thus "microbes" (and, subsequently, "microbiomes") always come into being for humans by way of scientific interlocutors and, in this way, have been enacted throughout history in tandem with the technologies developed to see them. In the first two decades of the twenty-first century, the development of metagenomics and associated data technologies completely transformed how scientists study microbes. By way of microscopy, laboratory cultures, and molecular biology, advances in high-throughput sequencing technology have revealed a previously unseen, massive microbial diversity on and in the human body, ushering in a new era for the study of microbial communities in environments and hosts. "Seeing" microbes through their genomes has made visible a tremendous number of formerly unknown organisms and has moved research away from conventional, hypothesis-driven science.

Metagenomics, also referred to as environmental genomics or community genomics, pools and studies the genomes of all the organisms in a community and all the functions encoded in the community's DNA.[34] Metagenomics allows for vast populations of microbes, previously unculturable outside their habitats, and functionally inseparable from their communities, to be identified through their DNA. It addresses a more ecological approach to understanding microbes within their environment, how they function, and their intricate evolutionary relationships. The methods of metagenomics include a series of experimental and computational approaches centered around shotgun sequencing, massively parallel sequencing, and bioinformatics. Genomic technologies and high-throughput sequencing such as those used in metagenomics have transformed what it means to be a biological scientist. Big data innovations have changed what is knowable and what is valued; how scientists conduct research; and, consequently, how they understand biology.[35]

The study of microbiomes pivots on the union of these new metagenomic technologies with the scientific practice of creating and using "germ-free" experimental animals, or gnotobiotics. Gnotobiotics and germ-free are not technically the same but are talked about interchangeably. *Gnotobiotic* refers to an experimental animal in which the only microorganisms present are known (*gnostos,* "known"; *bios,* "life"), whereas *germ-free* means completely without microbes. Specific pathogen-free animals, which have become a lab standard, can be seen as a universally used type of gnotobiology; they have the same undefined commensal microbes as conventional mice with a discrete set of pathogens filtered out. Human microbial ecology depends heavily on mouse models—the standard experimental protocol is to transplant the microbes from human samples into germ-free mice. Gnotobiotic technology has become a regular experimental tool as a way of isolating communities of microbes, studying interrelationships of microbes and their hosts, and monitoring external effects like diet and environment. These experiments are meant to capture an individual's microbial community at one moment in time and then replicate it in multiple mice to determine the degree to which phenotypes can be transmitted via microbiota, to make generalized human conclusions.

According to a survey done by fourteen scientific government agencies in 2016, most of the microbiome research in the United States (37 percent) was done on human microbiota, followed by nonhuman lab studies at 29 percent. Studies on soil, water, and atmospheric microbes collectively make up only 22 percent of all research.[36] In 2017, the NIH awarded $42 million to three centers for research on the human microbiome: the Human Genome Sequencing Center at Baylor College of Medicine, the Washington University Genome Sequencing Center, and the J. Craig Venter Institute. Undoubtedly, how microbes make up (and work within) human bodies is the primary scientific concern.

In the years I've been doing this work, subfields in anthropology, sociology, geography, public health, and philosophy of science have emerged, taking particular notice of microbial life.[37] At first, this new literature explored the social lives of microbes, a microbial ethos, and to varying degrees has called for a microbe-down reorientation of the hierarchy of life-forms. This was followed by many (many, many, many) articles asking what it means to be human in the age of the microbiome. Social scientists joined the conversation about the commensal biosocialities of human microbial ecology, and lines were drawn between the "perils and promises" of thinking with microbial communities as model ecosystems.[38] Many social science microbial perspectives were rooted in the work of Lynn Margulis, a revolutionary biologist whose concept of symbiogenesis emphasizes the genetically historical interdependence of all life-forms on earth and places microbes squarely at the origin of evolution, society, and life. Symbiogenesis is the emergence of a new organism (phenotype, organ, tissue, or organelle) from the interconnectivity of two separate organisms—symbiosis over time, a relationship between organisms that results in the evolutionary change of both[39]—or, as Donna Haraway puts it, evolution through the "long-lasting intimacy of strangers."[40] Haraway embraces Margulis's thinking, first in *When Species Meet*[41] and later in *Staying with the Trouble,* conceiving of microbes as companion species, messmates, tiny makers of our material and discursive selves that we become-with, or not at all.[42] Anthropologist Gísli Pálsson, in his volume with Tim Ingold, *Biosocial Becomings: Integrating Social and Biological Anthropology,* thinks of humans (by way of microbiomes) as aggregates of life-forms and the outcomes of ensembles of biosocial relations: "Humans may usefully be regarded as fluid beings, with flexible, porous boundaries; they are necessarily embedded in relations, neither purely biological nor purely social, which may be called biosocial; and their essence is best rendered as something constantly in the making and not as a fixed, context-independent species-being."[43] Pálsson goes on to argue that the human–microbe, biological–social assemblages affect the very way social scientists can conduct research and that "a radical separation between social and biological anthropology seems theoretically indefensible."[44]

Philosophers of science argue that taking more notice of microbes as the dominant life-form on the planet, both now and throughout evolutionary history, will transform some of the philosophy of biology's standard ideas on ontology, evolution, taxonomy, and biodiversity. Furthermore, metagenomics impacts how social scientists make sense of communities, interactions, and environments, shifting analyses of life from entities to processes.[45] Next, ethnographers like Alex Nading attempt to translate microbial thinking into methodology, raising questions about how to do a social study of the microbiome: "human–microbe relations can sometimes be measured numerically, but they cannot be fully explained with quantitative tools."[46] Nading compares microbiome work of U.S. scientists and Nicaraguan hygienists, suggesting different ways in which the cultural/interpretive evidence of paraethnography interfaces symbiotically with the quantitative/statistical evidence of bioscience. Microbiomes follow interspecies trajectories through the work of those like anthropologist John Hartigan, who theorizes culture and society across species lines, attempting to reverse the trajectory by which social theory is generated, eschewing the assumption that society and culture are solely about people.[47] Work by scholars in human geography engages with the growing scientific, popular, and policy interest in the microbiome. Geographer Jamie Lorimer's work on probiotic and "rewilding" therapies suggests new ways of thinking of policy, science, and microbiomes as tools for improving environmental and human health.[48] Academic research from groups like the Economic and Social Research Council–funded "Good Germs/Bad Germs" project at Oxford, the Canadian Institute for Advanced Research's "Humans and the Microbiome" program, the University of California Humanities Research Institute Microbiosocial Working Group, the Social Study of Microbes Group at the University of Helsinki, and the transdisciplinary Microbes and Social Equity Group out of the University of Maine explores the transformative potential of recent developments in metagenomics for developing new public understandings of the microbiome.

Since human microbiomes entangle microbial processes with intimate human social practices, tools for understanding human microbial ecology also entangle the life and social sciences. Slowly, the collaborative needs of human microbial ecology are resulting in real-world research projects. In 2019, a group including a microbial ecologist, a biological anthropologist, and an anthropologist-historian (who have been working on a cross-disciplinary project called "Afrobiota") proposed that "microbiome research should integrate multiple scales, levels of variability, and other disciplinary approaches to tackle questions spanning conditions from the laboratory to the field."[49] Another team consisting of scholars across disciplines suggests that there are "opportunities for integrating microbiology and social equity work through broadening education and training; diversifying research topics, methods, and perspectives; and advocating for evidence-based public policy that supports sustainable, equitable, and microbial wealth for all."[50] An immunobiologist and public health scholar published a paper in *Nature Reviews Immunology* about how poverty affects diet, diet affects the microbiome, and the microbiome affects chronic disease.[51] In 2020, an international working group of which I am a member put out a paper attempting to set the agenda for social science research on the microbiome.[52] Junior microbiome scientists are making moves toward integrating social science priorities into their research: molecular microbial ecologists Erin Eggleston and Mallory Choudoir work on connecting environmental microbiomes to social inequity across temporal and ecological scales,[53] and Katherine Amato's lab at Northwestern is investigating host–gut microbe dynamics in human populations around the world, focusing on microbial contributions to host nutrition during periods of reduced food availability or increased nutritional demands like pregnancy.[54] In 2021, I had a productive and generous writing collaboration with biologists Travis De Wolfe and María Rebolleda Gómez and geographer Mohammed Rafi Arefin in which we tackled racism in the microbiome.[55] This trend is an acknowledgment by biological and social scientists working together that microbes are essential to all forms of health and thus biosocial causes and outcomes must be studied by biosocial teams. This work illustrates the important concomitance of microbiome science with debates in the social sciences concerning materiality, biology, and nature. Increasingly, social scientists may be able to analyze and work with human microbial ecology, mindful that the old equation of bio plus social equals collaboration inevitably fails. Anthropological, public health, and biomedical perspectives converge to "make microbiomes," but with the important acknowledgment that we aren't all in the same shit together.

Microbiosocial Futures

The new tools and creative experimentalism of microbiome research open up discussions of the limits as well as the horizons of possibility; there are growing ethical, legal, and social challenges resulting from studies of human microbiota. Besides worries that not enough is understood about host–microbial and other interspecies interactions to avoid unforeseen adverse outcomes, microbiome scientists are extremely careful about making ambitious microbiota/health claims and urge caution against any prescriptive recommendations. Microbiologists and epidemiologists also warn against an overinvestment in the microbiome as the new biological panacea for all humanity's ills. Next-generation sequencing technology can enable the discernment of "metabolic networks" and reveal biochemical reactions, but "genomes are littered with clues both true and false, such as 'hypothetical proteins' and genes that are understood poorly or not at all, but could make for important differences in what metabolic networks do."[56] The vastness and complexity of microbiome networks and their interactions (including social and environmental ones) need to be characterized before causation can be determined. Professor of medicine (with a focus on the epidemiology of risk) J. Dennis Fortenberry calls for special attention to be paid to the sociopolitical nature of microbiome research—problems of race, gender, and class have certainly begun to arise. There are further concerns about the implications of using microbiota to profile, or forensically identify, individuals or populations—microbiome science ultimately reifying categories it had endeavored to disrupt. As Fortenberry posits, "the challenge to microbiome research is to translate its insights into better understanding of health disparities without depending on the validity of the categories on which disparities are based. In fact, the microbiome becomes a critical tool for understanding the pervasive influences of inequality that are the social, psycho-physiologic, and environmental contexts that link racial/ethnic categories and health."[57] Some from public health and the National Human Genome Research Institute have called for microbiome science to be used in an integrated health disparities research approach, requiring "the collaboration of investigators from multiple disciplines, including basic and computational scientists, clinicians, social and behavioral scientists, and epidemiologists. Each discipline plays a vital role in delineating the relevant factors that accumulate over the life course to influence disease risk and differences in health outcomes."[58] In this way, anthropology of microbes tries to work on what is hidden as well as what is revealed.

Medical ethics and public health scholars are concerned with dietary supplements resulting from human microbiome research, which open the door to what they call *commercialized intervention.* Already, microbiomes are being marketed as manipulable agents of health that can be controlled through the purchase of specially designed products, such as probiotics, personalized cosmetics, specially designed foods, and even breast milk supplements. This research direction raises questions of profit and unequal access to and distribution of technologies as well as the vulnerability of publics to "molecularized marketing."[59] Microbiome research seeks to find how microbes can be utilized, genetically altered, or even prescribed to effect durable changes in the environment, food, and bodies. As the biovalue of human microbiome research is in danger of being increasingly assimilated into capital value, public health interests, especially in the case of traditionally vulnerable populations (where disease burden and structural violence are greatest and a history of research exploitation persists) need to be carefully balanced with the scientific research, funder goals, and the industry marketplace.

As human microbial ecologists stand on the precipice of defining human health through relationships with our coevolved microbiota, work also emerges for scholars in the social sciences, public health, and bioethics, investigating the proprietary ownership of microbes and new necessities of biomolecular privacy required. We must think about how our genomic intertwining with our microbial partners changes practices of biomedicine and care. Reconsidering humans in the context of the microbial forces an interspecies perspective on a reluctant Anthropocene. Most importantly, we can use social science methodologies to elucidate the ethical and material conditions in which microbiomes come into being, to strive toward forging health intervention strategies that account for different relational ontologies between people and microbes. Unexpectedly, what emerged from my work with the Gordon Lab was an understanding not only of what the science is and is doing but of what anthropology is for me. All the involutions of *inter-* led me to the center: the inextricability of human bodies and lives from microbiome science.

Chapters

Gut Anthro uses a framework of shifting microbial ontologies to tell the coevolving stories of the social and biological sciences and also to situate physical (Gordon Lab, Dhaka, mouse guts, Bengali homes) and conceptual (anthropology, science, genomics, public health) spaces as coevolving sites. Human–microbe relationships can be equated to other binaries that have also been productively muddled, such as nature–culture, wholes–parts, anthropology–science, and nonhuman–human. When one starts to enlarge the scale of view, microbes multiply. Meaning and matter increase, as the ways in which microbes are shared, transported, and changed increase exponentially. Microbial scaling goes both in and out, big and small. What is at stake in these new relationships? The means through which human microbiota are studied and understood propels future research directions, raises bioethical issues (who owns your shit?), and complicates logics of funding (NIH, private foundations, major food manufacturers). This book explores how the study of human microbial ecology opens new ontological terrain for social scientists. Through microbes, life can be seen not as a property of individuals or humans but as one of coevolution, agency, and circulation. With an ethnographic focus on collaborative practice, this book takes microbes as actors, enacted differently in each chapter.

Chapter 1, "What We Talk About When We Talk About Collaboration," develops a descriptive analysis of what an interdisciplinary, collaborative anthropology of microbes looks like in practice. Just as humans and microbes are "becoming together," biological and social lives are coevolving entities, revealing in turn the interdependence and simultaneous coevolution of the fields of human microbial ecology and anthropology. This chapter describes my relationship to the Gordon Lab, outlines our collaborative work plan, and attends to the following questions: What are the terms through which microbes are becoming a research and social object? How does collaborating change outcomes of biomedical research and public health interventions that situate them? Are there limits to what we can learn from one other, and how do we work past these boundaries? Can disciplinary frictions be sidelined to devise real-world action that accounts for the historical and biological toll social violence takes on bodies? Can biomedical and socioeconomic interventions be cooperative and coprioritized?

My partnership with Dr. Gordon was predicated on a two-part agreement: first, that incorporating anthropological analyses into the design and interpretation of studies of human microbial ecology can provide scientists with crucial information about the specific social, political, and economic factors that shape human microbiomes, and second, that investigating microbes with ethnographic tools will provide anthropologists with new perspectives on the inextricability of human biology and social practices. Together we believed that negotiating the distinct and divergent methods, vocabularies, and conceptual categories that exist between anthropology and human microbial ecology is a timely and worthwhile challenge. Chapter 1 documents this ongoing cross-disciplinary dialogue, which explores new views of "humanness," "community," and "life" and, in the process, seeks to address major global health inequities. Pressing anthropological concerns (and coinciding disciplinary challenges) are emerging from scientific studies of the human microbiome. Carefully performed biosocial collaborations have the potential to transform the ways in which translational science is done, beyond microbiome-based interventions that try to circumvent sociomaterial vulnerabilities. With these efforts come unique opportunities for social and biological scientists to collectively face the concrete issues of health and illness, poverty and pain, that are embedded in human–microbe relationships.

Chapter 2, "How to Make a Microbiome," serves as an abbreviated but dense history of microbiology that lands the microbiome squarely in the lap of the social sciences. By tracing in tandem the parallel histories of microbes and scientific modes of seeing microbes, I look at how work across disciplines (including anthropology) has helped to produce the microbiome as an experimental object. This chapter investigates how microbiome scientists negotiate intellectual and social barriers and how their work produced something called a microbiome. New ways of understanding microbial life are having a revolutionary effect on frameworks of evolution, taxonomy, and multicellularity. For every topic in the biological sciences, from neuroplasticity to cancer, there are important associated microbiota. In turn, microbial concerns emerge for the social sciences, humanities, and philosophy along themes of life, relationality, health, science, and embodiment.

Chapter 3, "Microbiokinships," asserts that microbes are kin—kin that are made of and making environments, across generations. Microbes coevolve with humans, and microbial populations in human bodies are determined by environments and exposures, including family, food and place, health care, race and gender inequities, and toxic pollution. Microbiomes are transgenerational links, disarrangements between different bodies and the outside world. Microbial kin evokes Indigenous formulations that necessitate reciprocal, ethical, and environmental accountability to more-than-human relations. Environments can be the natural world plus toxins; they can be genetic landscapes that span generations and geography. Environments can be global, extraterrestrial, and cellular. Environments can be people.[60] They can be uterine, chromosomal, gut, vaginal, atmospheric. There is no doubt we are permeable, whatever, whoever, "we" are. As long as we talk about exposures, there will be something out and something in, someone to be exposed and something to be exposed to. Microbes cross borders; they can constitute the environment and can alter it. Microbes in human bodies are an intersection of fleshy materiality (genes, birthing, bacteria) and social intimacy (bathing, breastfeeding, affection); they circulate and confound the inside of bodies, the outside world, and back again. Microbes are in and out, human and nonhuman, simultaneously environment and body. What we mean by environments and kinship may be refigured through a focus on microbes.

Chapter 4, "Malnutrition Futures," focuses on the close tie between understanding the microbiome and associated genomic technologies. In recent years, next-generation genome sequencing has provided new ways of seeing the hundreds of trillions of bacteria in human bodies. Through "datafication," new forms of value are emerging from microbial genomic information—implicating microbes as powerful agents of nutritional status. Scientists use metagenomic data to evaluate how food, environment, and genes affect human gut microbiota and how those microbes simultaneously affect human health. This chapter looks at the social and material conditions of the Bangladeshi women and children enrolled in the microbiome study and how microbes are datafied in order to draw a causal relationship between microbial populations and undernutrition in human hosts. How is malnutrition lived for urban Bangladeshis, and how is malnutrition studied through metagenomics? What categories of undernutrition and bodily health emerge as big data becomes a tool for nutrition science? Practices in homes and communities in Dhaka and how malnutrition is studied in the lab are both choreographed with data-producing microbiome technologies in the quest for translational health care strategies that treat childhood maladies in developing countries.

Chapter 5, "Ghosting Race," looks across microbiome research and asks, What is race doing in these studies? Why is it there, and how is it functioning? Human microbiome studies gesture toward the postracial aspirations of personalized medicine—characterizing states of human health and illness microbially. By viewing humans as "supraorganisms" made up of millions of microbial partners, some microbiome science seems to disrupt binding historical categories often grounded in racist biology, allowing interspeciality to supersede race. But inevitably, unexamined categories of race and ethnicity surface in a myriad of studies on microbiota. I examine this research to argue that microbial differences need a social science perspective—to investigate how microbiomes and race are entangled embodiments of the social, environmental, and biological. Ultimately, transdisciplinary collaboration is required to address racial health disparities in microbiome research without reifying race as a straightforward biological or social designation.

The book's conclusion discusses next steps for an anthropology of microbes—a complete picture of all the shit: the forestalling and failures, the anguish and small victories of my experimental entanglement with the Gordon Lab. I venture a disciplinary, intellectual, and practical cost–benefit analysis to try to parse out whether it was worth it in the end. I have learned a lot since starting this project in 2009. I am not the same person; the citations I made, the commitments I had, the way I did fieldwork, were different. What I read and what I care about now are different than in 2009, 2012, 2016. I can't rewrite the entire thing now. Evolving is not smooth, effortless, or graceful. Things are lost; things are damaged. You may not recognize yourself when you come out on the other side of the shit; you may not even be a singular *you* anymore. What anthropologists are and should be, what humans are and should be, is called into question through our life-and-death relationships with microbes. We can't be safe and complacent within our familiar categories of academic research fields or of species. The anthropological study of microbiomes compels ethical commitments that require radical, actionable changes that cannot be contained by these boundaries.

There is one field office and three local icddr,b nutrition centers like this one in Mirpur camps B, C, and D. In each center, one field research assistant supports supplementary feeding and provides health and nutrition education, and four field research assistants do data collection. Sixty mothers come to these local centers daily for nutritional supplements (*pushti* packets).

ONE

What We Talk About When We Talk About Collaboration

> This is a field that needs cultural anthropologists—people who can view how our cultural traditions influence our lifestyles and shape choices that in turn affect our microbial ecology. Given its "newness," metagenomics provides a way for anthropologists to study, in real time, how the field influences the scientists involved in its evolution (i.e., how do we view ourselves as a result of our discoveries? How do our discoveries affect our narratives and our lifestyle choices?). . . . How do this field and the participants' cultural traditions interact to affect their perceptions about forces that affect their destinies, or their connections to one another within the context of a family or community? The union of metagenomics and cultural anthropology can reveal how cultural traditions, such as how infants are handled and cared for in early life, influence the flow of microbes between generations of humans, thereby shaping the physiologic and genetic features of a kinship.
>
> —from a letter of support Dr. Gordon wrote for a National Science Foundation (NSF) Doctoral Dissertation Research Improvement Grant I received before I began my fieldwork in his lab

An anthropology of microbes is a precarious house, built on the shifting ground of collaborative promise. As we rethink objects and subjects of study, and our reframed relationships to them, this chapter develops a descriptive analysis of what my transdisciplinary, multisited, collaborative anthropology of microbes looked like in practice. Just as humans and microbes are "becoming together," biological and social lives are coevolving entities, revealing in turn the interdependence and simultaneous coevolution of the fields of human microbial ecology and anthropology. In my work, anthropology of microbes required collaborative efforts on the part of bacteria, babies, mothers, field research assistants, Bangladeshi scientists, scientists in the Gordon Lab, and global health funders. These endeavors pulled together nodes that were mutually beneficial and disjunctive, unproductive while also indescribably important. In a time when social, biological, ecological, and political approaches to questions about the world refuse the disciplinary silos they've been assigned, this kind of collaborative work is critical—but also deeply difficult and sometimes painful.

This chapter describes my relationship to the Gordon Lab, outlines our collaborative work plan, and tries to untangle and make sense of the complex relationships that come from collaborating. I concentrate on the following questions: Does collaborating change outcomes of biomedical research and the public health interventions that situate them? Are there limits to what we can learn from one other; are there bridges that cannot be crossed? Can disciplinary frictions and epistemological agonizing be sidelined to devise real-world action that accounts for the historical and biological toll social violence takes on bodies? Where are the ontologies of the anthropologist and the human microbial ecologist at stake in their relating?[1] Can biomedical and socioeconomic interventions be cooperative and coprioritized? Are personal and scholarly injuries inevitable, and are there wounds that cannot be healed?

Jeff Gordon's initial, enthusiastic reception of me and my work developed into concrete, material, and intellectual bonds, and in a formal offer letter, he invited me to spend a year in the lab: "I was delighted to read your proposal and feel that it certainly meets my aspirations for your project. As we discussed, this project will be conducted in St. Louis, Malawi[2] and Bangladesh, beginning September 1, 2010, and proceeding through August 31, 2011.

Therefore, I am delighted to formally offer you the following positions: Visiting Researcher with Student Status. I understand from our discussions that the New School, where you are enrolled as Ph.D. student in Anthropology, fully supports this arrangement, which is defined as 'performing field work for your thesis.'"[3] I was honored and astonished to be offered such a position and taken aback by Dr. Gordon's welcoming generosity. I felt so lucky, and also very scared. I was an anthropologist-in-training; was I capable of doing all that the lab might expect of me? I felt I was representing all of anthropology and anthropology's capacity to contribute meaningfully to microbiome science; what if I fucked it up?

My entanglements were knotty, and many. I worked for (and was paid by) the Gordon Lab, doing anthropological and ethnographic work for Dr. Gordon[4] in support of the lab's objectives in St. Louis and Dhaka. I was on the Gordon Lab payroll for over a year. I was also a PhD candidate pursuing my own dissertation research, funded by my own personal NSF grant. Furthermore, my partner, Joe (also a trained academic anthropologist), was hired by the lab, and we traveled, worked, and wrote for Dr. Gordon together—our family's economic viability, ability to live in the same city, and health insurance all dependent on our lab employment. I had privileged access to data, experiments, and field sites and at the same time was routinely viewed by lab members as a second-order participant-observer. Early into my fieldwork, one postdoctoral fellow confronted me in a lab meeting in front of all the members of the Gordon Lab, expressing concern about his privacy and confidentiality and objecting to speaking "on the record." Some lab members and staff were generous with their time and thoughts, open and kind, and became excellent interlocutors and friends. Other people avoided me for the better part of a year, dubious or perhaps afraid that I was going to write a lab exposé that would imperil their future scientific careers. I was a highly educated thirtysomething white person in a lab primarily populated with other white, educated thirtysomethings, which at times felt disciplinarily like *studying up*[5] but felt socially like *studying laterally*. Formally, I was fully incorporated as a lab member, listed on the lab website, with an official university ID and Wash U email address, but my yearlong presence was surrounded by a cloud of perplexity and suspicion. At times, I was as confused as my lab colleagues about whether my position was inside or out.

Initially, I didn't expect my work to become so integrally intertwined with the lab's work. I wrote in my original research proposal that I intended to "investigate," "observe," and "critically examine" how the Gordon Lab was studying microbes. Instead, what emerged was a momentous intellectual and life course: attempting to create a productive interaction between anthropology and human microbial ecology. I became concerned with how microbial perspectives are playing out in the study of human health, what types of global health interventions are possible and promoted (and on what types of bodies), and how such work might change the equation in the practice of social science. The idea of forging a new kind of anthropology was a thread that ran throughout my conversations with Dr. Gordon. He told me, "My philosophy is not to ask permission of the establishment. It's to present what you've got and say 'here it is.'" He talked about science like anthropologists talk about science—interdependent, networked, social, complicated. Dr. Gordon believed that "a melding of the field of anthropology and human microbial ecology research is absolutely essential for the design and interpretation of studies" such as the ones he and his students were conducting, in diverse socioeconomic and cultural populations across many countries and continents. He was also "very intrigued by the question of how the lab's work is affecting how members perceive themselves and how it impacts their daily narrative." These ideas were strongly aligned with the kinds of questions I wanted to ask of microbiome science. I was surprised and excited to encounter these social science perspectives from a prominent biological scientist, and these perspectives bonded us and fortified our collaboration. Being a lab member and watching the lab work get done also tapped into my deep discontent with the problems of anthropology. Anthropologists often think of themselves as the peoples' social scientists yet can be willfully ignorant of all the race/gender/class/ability power structures tenaciously at work within anthropology. Fear of science, terror of acquiescing to scientific authority, has made some anthropology antiscience and antimaterialist to a dangerous fault. When I applied for a grant from the Wenner-Gren Foundation, a preeminent disciplinary institution in the advancement of anthropological knowledge, one of my reviewers wrote, "This project is not anthropology." But Dr. Gordon made me want *to do anthropology better,* to figure out a better kind of anthropology. This task was an ongoing, backbreaking thing. When I checked in with my anthropology faculty, they were worried: "they expect little ol' you to do *all* that?"[6] I was afraid of becoming the (unqualified) ethical and social spokesperson for all microbiome work, but this was not necessarily because of pressure from Dr. Gordon. He often expressed how grateful he was for whatever contribution Joe and I could make.

Dr. Gordon and I devised a two-part, aspirationally reciprocal agreement that would guide our partnership: we agreed, first, that incorporating anthropological analyses into the design and interpretation of human microbial ecology could provide scientists with crucial information about the specific social, political, and economic circumstances that shape human microbiomes and, second, that ethnographies of microbes would provide anthropologists with new perspectives on the inextricability of human biology and social practice. We acknowledged that we had distinct and divergent disciplinary methods, vocabularies, and conceptual categories (though at the beginning I don't think either one of us realized the magnitude of that divergence), but together we believed that nothing was more important than facing that challenge. This chapter documents the painstaking assembly of this anthropology of microbes and tracks a cross-disciplinary dialogue created in the service of cooperatively addressing major global health inequities. I address how ethnography contributed to, interfered with, and diverged from the microbial ecology research (and vice versa). Dr. Gordon and I ardently shared big ideas, but how we would materialize and practice them was continuously indeterminate. I swung between spaces of "too critical" for the scientists and "not critical enough" for the anthropologists. After I left St. Louis, things became even more complicated, and the more intellectual, temporal, and geographical distance I put between myself and the lab, the more things broke down.

Collaboration, Ad Nauseam

By now, *collaboration* is so buzzy a buzzword that it is hard not to cringe from its overuse. Perhaps the only word used more in social sciences of biomedicine and science than *collaboration* is *interdisciplinarity.* Full volumes are dedicated to analyzing the practice of interdisciplinary collaborations—*Rethinking Interdisciplinarity across the Social Sciences and Neurosciences,*[7] *Investigating Interdisciplinary Collaboration,*[8] and *Sustaining Interdisciplinary Collaboration: A Guide for the Academy,*[9] to name a few—and the journal *Nature* published a special "Interdisciplinary" issue in 2015 in which one author claimed, "To solve the grand challenges facing society—energy, water, climate, food, health—scientists and social scientists must work together."[10] New bioscientific models, including neuroplasticity, systems biology, epigenetics, nonlinear modeling, and human microbiome research, focus on a *flexible biology.* This turn toward a nonrigid, mutable biology inextricable from the social and technological appears to be a significant inroad for social science.[11] Angela Willey has examined this classically Kuhnian paradigm shift as a move away from traditional essentialist science, yet warns against leaving the science uninterrogated:

> The new science, that begs a new feminist disposition, is a science not hell-bent on imperial domination of the world through exhaustive mapping of its constitutive parts, but rather a science that acknowledges the coevolution of the "natural" and the "cultural," the "human" and "non-human," the "living" and "non-living." A science beyond epistemological reproach, because it has, in theory, ceded ambitions to harness the essence of nature and understand the social world in terms of said essence to a naïve scientific past.[12]

Other work is being done to problematize this focus on plasticity, not as all potential and possibility, but as a strategy for understanding new modes of kinship, reproduction, cells, biological sex, immunity, and hormones.[13] Largely, social scientists view this sea change in the life sciences as a signal that "the time has come to reposition this historical legacy and to move beyond the acrimonious controversies that have characterized twentieth-century thought as it traversed the biology/society border."[14] Likewise, humans and microbes become together, modeling a coevolution of the biological and social in the social sciences, not just ideationally, but materially.

Since the early 2000s, turns in STS and the social sciences have moved toward ontologies and new materialisms. On the simplest level, ontology is concerned with worlds rather than worldviews. New materialism seeks to escape the classic problematic dualisms of subject–object, material–cultural, and nature–nurture by seeing the meaning and matter of the world as inextricable, reclaiming the material world that has been elided by social constructionism.[15] Both movements have strong roots in feminist theory, and both have been embraced and interrogated by feminist anthropology. Both have pushed the social analysis of feminist anthropology toward reevaluating relationships with science, nonhumans, and our own disciplinary commitments. These turns have also faced criticisms; there is concern about neglecting anthropological obligations to human cultural lives, worry about the foreclosure of the political, and accusations that science is being taken at face value. Some also ask for the deconstruction of the monolithic concepts of "Science" and "Feminism" and to explore objects of feminist thought and science: "what worlds do their varied knowledge projects come out of and lend themselves to materializing?"[16] I'm proposing here that I am willing to commit to a new engagement with microbiome scientists in pursuit of a future science that doesn't elide the past but allows us to "live with its ghosts," as Subramaniam suggests as she "embraces the tensions between being a feminist (with all the commitments that entails) and being a feminist who recognizes the collusion between scientific knowledge and categories of difference."[17]

Working *with* science can push feminist anthropologists to go beyond categories of biological and social, to see these parts of life as intra-acting. By coming back to biology and materiality and considering those as agentive within feminist analysis, these theories of action allow anthropologists to account for the complex interactions between the bodily, worldly, fleshy, intimate, emotional, cognitive, and metaphysical and help scientists, engineers, and technologists to carry out ethical projects, with mutual engagement and credit, with what Subramaniam and Willey promote as "disciplinary promiscuity."[18] Although the legacy of feminist science studies is to question the methods, objects, and objectivity of science, feminist STS since the early 2000s has elaborated new perspectives to reimagine nature, biology, and matter (though not without debate).[19] How, then, can feminist theory and science be brought into a more fruitful alliance? Are there ethical commitments that extend beyond disciplinary boundaries? As Subramaniam says, referring to Stengers, "the challenge then is not a pointless battle between the irreconcilable frames of objectivity and social construction, but how to better understand the tensions between objectivity and belief as a necessary part of science and as central to the practice of science."[20] In her book *Gut Feminism,* Australian feminist theorist Elizabeth A. Wilson laments the traction feminist theories of the body and embodiment have gained by eschewing biology. She works through an analysis of the pharmacokinetics of antidepressants and placebos, suicidal ideation and bulimia—taking the gut as a site of metabolization and politics. Starting with the premise that the gut is an organ of the mind (is always minded), Wilson uses new theories of biopsychiatry to try to productively process pharmaceutical data—not to critically deconstruct them, but to use them. She explores the organic interior of the gut and its psychic nature, rejecting the notion that stomach and mood were never not coentangled, that psyche and soma were ever not coevolved. How pills, synapses, gag reflexes, and depression transform and interact in the gut is contingent on emotion and environment. She suggests a generous engagement with biology that has the potential to reanimate and transform the foundations of feminist theory—indeed, the call for feminist theorists to develop a conceptual tool kit in which we can take biomedical data seriously, but not literally. Yet Wilson and others' most powerful provocations are in recognizing just how crucial it is that we start trying to figure out "pathways by which biological data can become critically mobile"[21] and ways in which feminist theories can shape what science could be.

Many have written about disarticulating biology and culture and, recently, about rematerializing matter. But so far, difficulties remain in getting different disciplinary knowledges or different data to actually transact, what Wilson describes at one point as "a way of articulating those nonlinearities with each other . . . realms both align and dissociate, how they are antagonistically attached,"[22] bringing to mind Strathern, Mol, and Barad. Feminist anthropology's extensive study of the sociocultural and political implications of biological sciences and medicine, as well as its rich history of ethnographic evidence, positions it well to take up this charge. Some anthropology has always been close kin with feminist STS, with shared commitments to studying how power, identity, biology, and technology are tangled up with socioeconomic and cultural life. Ethnography is especially suited to on-the-ground science studies research, not unquestioningly accepting scientific truth but striving to take material biology seriously, compelled by practices, outcomes, and action. Willey introduces a useful conceptual resource for feminist scholars studying science through *biopossibility,* a "species- and context-specific capacity" that helps hold "the material-discursive conditions of scientific knowledge production and the materialization of bodies in the same frame."[23] One possible trajectory for exploring *biopossibilities,* in order to bring together ontology, new materialism, science studies, and feminist anthropology, is to work collaboratively and ethnographically with biological scientists.

Elizabeth Roberts has brilliantly outlined a functional methodology: *bioethnography,* combining biological samples and data with ethnographic details and environments to produce knowledge not independently possible, where neither type of information is privileged or discrete.[24] Ethnography as an analytical approach can be a critical first- and second-order methodology in investigations of how social determinants of health are also biological—for instance, how social vulnerabilities to poverty and race can be enacted on the body through biological illnesses like lead poisoning, asthma, and diabetes. As Roberts has pointed out:

> contemporary feminist anthropologists, enmeshed in ongoing technological ambivalence, base their arguments not on essentialist nature as their guide, but within an expanding approach that posits nature as constructed, and medicine, science, and biotechnologies as a set of practices and realities specifically embedded in particular material circumstances saturated with power relations that shape how differently situated people will use them.[25]

When I began my work in the Gordon Lab I didn't account for the power of seniority in one's field—how scholars like Roberts, Callard and Fitzgerald, and Fortun and Fortun were able to construct and negotiate collaborations with scientists because of their own relative power.[26] As an extremely junior, not-yet-professionalized anthropologist, I was fueled by hope and too heedless of my inferiority. I wanted to realize biopossibility, enact bioethnography, provide the Gordon Lab scientists with qualitative ethnographic data about living conditions, social networks, and daily practices, contributing to the microbiome work. I wanted to provide a more socially meaningful context through which this science could act, and it seemed that in Dr. Gordon, I had the perfect partner.

Nodes

Over two years (2008–10), three groups of babies aged newborn to eighteen months participated in the Gordon Lab Microbiome Discovery Project, an unofficial arm of Mal-ED. Two hundred were in a birth cohort, mothers enrolled prenatally and the children born into the study, and six hundred case-control children were enrolled in infancy, determined to be moderately or severely malnourished before being accepted into the study. "Case" children have −2 Z-score and "control" children a −1 Z-score. The Z-score classification system interprets weight-for-height, height-for-age, and weight-for-age. The Z-score system expresses the anthropometric value as a number of standard deviations or Z-scores below or above the reference mean or median value.

Mal-ED uses guidelines laid out in the World Health Organization (WHO) Global Database on Child Growth and Malnutrition, which includes the Z-score system. While widely recognized as the best system for analysis and presentation of anthropometric data for population-based assessment, the Z-score system has been reluctantly adopted at the individual level. Since the 1990s, Z-score classification has been challenged; it fails to be an accurate culturally or geographically relative indicator of child growth.[27]

The birth cohort in Mirpur consisted of children whose mothers were enrolled while pregnant. These mothers received extensive prenatal care; were given education on healthy eating, breastfeeding, and hygiene; and gave birth at BRAC Health Centres rather than at home, which was the local convention. Previously the Bangladesh Rehabilitation Assistance Committee, then the Bangladesh Rural Advancement Committee, and currently representing no acronym, BRAC is the largest development NGO in the world, working in areas of public health, education, and economy. BRAC banks (providing microcredit),[28] handicraft stores (selling goods locally made by Bangladeshi women), and health centers are ubiquitous in Dhaka. The case-control children in the study were given macro- and micronutrient supplements as well as therapeutic foods. All the families received free health care in exchange for their participation. Each family was visited regularly by local field research assistants (FRAS), who collected socioeconomic, surveillance, and food recall data. Sample collection and surveillance of cohort children began within seventy-two hours of birth—meconium, the first stool of the babies' lives, is the first sample they submit. It is their first bodily emission and their entry into the study.[29] Blood, urine, and fecal samples were collected regularly sent to icddr,b and then to the Gordon Lab for genetic analysis. The Bangladeshi feces were also used to "colonize" or "humanize" germ-free mice; these mice were then used in various diet experiments to see how the microbiota would respond to different foods and perturbations.[30] To "humanize" a germ-free mouse is to insert microbes from a human donor into an animal who has been bred and reared in sterile conditions, with no microbes of its own. Isolated communities of microbes can be introduced, and external effects, such as diet and environment, are monitored.

Fecal microbiota samples obtained from Bangladeshi children were introduced into the guts of germ-free mice.[31] These mouse recipients of human gut communities lived in plastic bubble isolators completely protected from any external environmental microbes—the only microbes in their bodies were human ones. The language of proprietary, species-specific microbes gets a bit tricky, for instance, "human microbes" versus "mouse microbes." This allowed members of the lab to learn how many physiologic, metabolic, or immune features of the human microbiota donors could be transmitted to recipient mice and thereby begin to establish causal relationships between human gut microbes and health status. The mice were given different diets representative of the microbiota donors (e.g., a "Bangladeshi diet," "Malawi diet," or "Western diet"), and then the microbial communities were analyzed using high-throughput culture-independent (metagenomic) methods. Lab members believed that this metagenomic data could be used to understand the origins and manifestations of disease in the very human population whose gut microbial communities were being used to generate these mouse models. Much like Nicole Nelson's behavioral genetics scientists in her book *Model Behavior*,[32] the Gordon Lab scientists were perfectly aware of the shortcomings of these models. They knew their data were partial and provisional, recognizing "this paradoxical nearness and distance between the mouse and the human, the need to pull them together while maintaining an awareness of their fundamental differences."[33] Ironically, humanizing mice reduced microbial relationships to their biologic components, looking primarily at the chemical compositions of food and functional microbial interactions with nutrients. And in the process, *humanizing* a mouse means to make its gut a microbial simulacrum of a human gut. To be human means to be microbe.

In this project, conditions of ethical, technological, and material possibility are altered as microbes move and are transformed through a disparate yet functional collaborative network. Nodes of this network include Dhaka markets and foods, Bangladeshi neighborhoods and bodies, local practices of childcare and hygiene, the Gordon Lab academic research facility, metagenomic data, germ-free mice in germ-free isolators, experimental diets, and the world's largest private philanthropic foundation. I followed the chain of transformations between two ends of this network, ends that were themselves transforming: the Gordon Lab and the Bangladeshi community of Mirpur 11.

Dr. Gordon began his career as a physician—after completing his clinical training in internal medicine and gastroenterology and doing a postdoctoral fellowship at the NIH, he joined the faculty at Washington University, where he has spent his entire career. He has been a member of the Departments of Medicine and Biological Chemistry, the head of the Department of Molecular Biology and Pharmacology, and, since the early 2000s, a founding director of the university's interdepartmental Center for Genome Sciences & Systems Biology (CGS). In 2017, CGS received $10 million from the Harry Edison Foundation and was renamed the Edison Family Center for Genome Sciences & Systems Biology. In 2022, the CGS housed ten labs, including Dr. Gordon's. When Dr. Gordon tells the thirty-year history of his lab, he tells it through people—not instruments, not even microbes, but through the genealogy of discoveries of various lab members. Some of the research staff have been with Dr. Gordon for more than thirty years—a testament to his enduring commitment to individuals and partnerships.[34]

The Gordon Lab, founded in the late 1980s, was originally focused on using transgenic and genetic mosaic mouse models to characterize how developmental stage-specific, cell lineage–specific, and spatial patterns of gene expression were regulated in the gut's continuously renewing epithelium. This work led to conclusions that gut epithelial cell differentiation is critically shaped by environmental cues, and the lab turned to the microbiota, wondering whether differences in epithelial gene expression along the length of the gut reflected a mutually beneficial crosstalk between microbes and host. What followed was a series of cumulative discoveries: the idea that a gut microbe could direct its host to manufacture a nutrient source; a marriage of gnotobiotics and functional genomics; the importance of early environmental exposures in defining gut microbial community structure; techniques for culturing the majority of bacterial taxa from previously frozen human fecal microbiota samples and transplanting into gnotobiotic mice to define the effects of diet manipulations; methods for transplanting and reliably replicating phenotypes from a given human donor in gnotobiotic mouse recipients; and, ultimately, a causal relationship between the gut microbiota/microbiome and a form of malnutrition that the lab referred to as *kwashiorkor* (the language shifted to *severe acute malnutrition* around 2014). Dr. Gordon told this story of his lab in a speech accepting the Robert Koch Prize in 2013. The Koch Prize is awarded for major advances in the biomedical sciences, particularly in the fields of microbiology and immunology, and is widely regarded as the leading international scientific prize in microbiology, often a stepping-stone to the Nobel Prize for microbiologists and immunologists.[35] Obviously, the Gordon Lab work continues to prodigiously progress (on my first visit to the Gordon Lab in 2009, I was told that they produced a paper a month, unlike other labs that publish one paper a year), but I will try to limit what I discuss here to the work being done during my collaboration with Dr. Gordon over 2009–12.

The Bangladesh samples became the next chapter in this story of microbial malnutrition—the 1,892 frozen stool samples, taken from 103 Mirpur cohort study children from birth to two years old, were pulverized, and the remaining cells underwent a process of purification and extraction. The resulting DNA was sequenced deeply, to produce 454 data sets, 242 shotgun-sequenced gene data sets. PhD student Brian Muegee found that between zero and twenty-four months, there was a total replacement of microbial communities, that age and diet are confounded, and that both are significantly related to changes within a child. Diet shifts are less important than aging because total metabolism changes with age. Although average taxonomic changes are smooth with age, each child is a complicated pattern of microbiota maturation—meaning that malnourished children seemed to have microbes more similar to younger children than children their own ages. This theme of microbiota "immaturity," and how to alter this immaturity, would become a central focus in the Gordon Lab approach to the malnutrition problem.[36]

My own work for the lab was hard to define concretely, hard to keep still. Starting in August 2010, I sat waiting for many days in my designated office, two doors down from Dr. Gordon's. Because there were no windows, he quickly equipped me with a digital picture frame to show a slideshow of landscapes. He instructed office assistants to buy whatever books I requested to build an anthropology of food library. He asked me to start a wiki, collecting information on Bangladeshi food practices. He was *so busy* that we did not have regularly scheduled meetings; he would just periodically whoosh into my office, a force of ideas and palpable enthusiasm. Dr. Gordon and I had many, many conversations; I made countless unsuccessful attempts to set the parameters of expectation, to define what I wanted to do, and to find out if doing it was possible. Combing through my two hundred typed pages of single-spaced field notes, I realized that I spent more than a year, from my first week to my last, struggling with the same issues. These notes also reveal what I thought anthropology should look like and what I thought the "right" way of doing fieldwork was:

> August 2010
>
> It's so unclear what Dr. Gordon wants from us—independently or together, that it is making it very difficult to figure out what we should be doing. Again, I know this is our first week, but I feel like it is a crucial time to be establishing our goals, boundaries, and how things are going to be in the coming year. I want to still do my own thing—but not entirely. Complicated. There's so much you can't know until it happens. No one in the lab understands anthropology.
>
> September 2011
>
> As awesome as it is that JG has extended my work with the lab, and I will have an income until the baby is born, I am dreading (a little) the fact that I will not be free, that I will still be indebted to JG, and working for him, and not able to fully commit myself to focusing on my own writing.

It seemed to me Dr. Gordon was more focused on the Dhaka data set, in hindsight perhaps because it was more valuable to the work being done in the lab; an anthropological analysis of his lab members wasn't likely to be added to their publications roster. Competing disciplinary alliances and personal and professional gratitude were so snarled in my research, writing, and life that I could not untangle them. One thing that was clear from the outset was that Joe and I would travel to the global lab field sites, the place where our expertise made the most sense and impact. The first Bangladesh trip happened suddenly in November 2010, meant to be a scouting mission of sorts. That was followed by a longer, six-week trip in spring 2011. Before I left for Dhaka, I was tasked with addressing the question, what can we say about family structure, living conditions, food habits/food distribution, childcare, and maternal/prenatal health? I was asked to photographically document conditions of sanitation, food, homes, and evidence of "westernization."[37] I devised three sets of questions to be approved by Dr. Gordon and the several institutional review boards (IRBs) about households and families, food acquisition and preparation, enrollment in the microbiome study, and health-seeking behaviors. In my slightly separate but not unrelated "own" work, I attempted to trace the internal complexities of how a global health project is defined and enacted, learning about the ontological navigation the Gordon Lab members performed through their research and their own conceptualizations of microbes.

During 2010–11, at least six of twenty Gordon Lab members were working specifically on malnutrition and/or obesity, and three were dealing directly with the samples from Dhaka.[38] While one postdoc was focused on cholera and how enteropathogens (organisms that cause intestinal disease) and the subsequent introduction of probiotics (beneficial microorganisms consumed for presumed health benefits) affect gut microbiota, I focused on two graduate student projects: Brian's, which was analyzing the Dhaka samples, and another conducting diet experiments using "Bangladeshi" food and germ-free mice humanized with Bangladeshi microbiota. I was interested in these studies because they were analyzing microbial material from babies in a specific

Mirpur homes are usually constructed from sheets of metal, bamboo, cloth, plastic bags, and sometimes brick.

geographical locale, with a specifically defined health and nutritional status. These studies involved human beings I would be able to talk to at a node in the network I could travel to and that I could observe. I wanted to find out where the shit came from—what the conditions were in the place it was produced and for the people producing it. I followed these samples to their origin, as they were made by human and microbial bodies, as they arrived in the lab, and as they were translated and transformed. Further, I was interested in how Bangladeshi diets were represented in mouse experiments and how dynamic human eating practices would work their ways into the scientific practice, or the ways they would be left out. Following Lock and Nguyen[39] and Koch,[40] I wanted to test the indivisibility of human activity and material environments through time and space; I wanted to take local biologies to the microbial level and find out how those local microbiologies were manifesting on the ground.

Urban Bangladeshis in the Gordon Lab study were subject to all the bodily, economic, and social risks poor and unprotected citizens participating in scientific research and health intervention programs face. It didn't seem that any of the American scientists involved were asking questions about equity or global, racial, or income disparity; none of them questioned why these bodies were required for this study. To address this, I wanted to shine ethnography's "empirical lantern"[41] on the practical, intimate realities of these families, using Paul Farmer's methodologies to attend to structural violence and corresponding issues like malnutrition from both proximal (potential personalized probiotics) and distal locations (like community-based clean water initiatives).

Methods

During my time in Bangladesh, I spoke with icddr,b FRAs, senior scientific staff in the Nutrition Programme, and nine families across four blocks, or community camps, in Mirpur, conducting intensive interviews and participant observation. I spoke primarily with mothers, but I also interviewed other extended family and household members. I visited each family over the course of a month and a half, accompanying mothers as they cooked, cleaned, and traveled to market and to the icddr,b nutrition centers. Five babies belonged to the birth cohort (two hundred families who were enrolled in the study while the mothers were pregnant), two were case children, and two were controls (of the five hundred case-control families who entered the study with a diagnosis of malnutrition). These babies and mothers were Rasel and Chandni; Zahir and Nipa; Reshmi and Mithee; Prianka and

Maliha; and Sadia and Sumi; two case children and mothers, Kanta and Rimi and Pavi and Mita; and two control children and mothers, Ashik and Anjana and Dev and Hamima.[42] Three FRAS, Roji, Sunita, and Zarat, chose the families for me based on the mothers' perceived willingness to discuss their experiences in the study and their comfort with the observation of daily household practices. All informal interviews and collective discussions took place in subjects' homes, in neighborhood streets, and in markets.

To conduct these interviews, I relied on the assistance of a local translator, Israt Akanda, a field research assistant with a degree in public health, and a native Bangla speaker. My access to the Mirpur families was provided by icddr,b because of my membership in the Gordon Lab, a factor that I took seriously and never for granted. This was a major translational health study, spanning continents and being performed by the top scientists in the field. There is no doubt I never would have been permitted access to the hospital or study subjects on my own. This is the double-edged sword of collaboration: the doubt you feel in the value of your own work, and the constant reminders that without your scientific partner, there is no access, no data, no project.

I tried to be transparent with mothers and scientists alike about my ethnographic research and what I hoped to investigate. I asked these mothers about their participation in the biomedical study, their understanding of the purpose, what they expected from the outcomes, and how their lives had been affected. I interviewed them about nutrition, hygiene, and health-seeking behaviors. They discussed food acquisition, preparation, and preferences with me; we talked about issues of childcare, family structure, and kinship. These women represented many diverse aspects of Mirpur; they were different ages, came from different places in Bangladesh, and had various opinions about the state of hygiene, health, and nutrition in their community. Many of them thought I was a medical doctor despite the many protests and explanations I beseeched Israt to relay.

On the other end of my field network, I spent eight hours a day in the Gordon Lab for one year: attending lab meetings; observing diet experiments in progress; watching bench science, DNA and RNA extraction, sequencing runs, and data analysis. I spoke informally with scientists daily about their work and the larger scientific and philosophical issues surrounding human microbial ecology. At times I was required to present my own work to the lab, explaining and accounting for what anthropological ethnography had to contribute both to the Gordon Lab's scientific work and to the world at large. After I returned to New York from St. Louis in late 2011, I continued to write with Dr. Gordon—our article in the *Proceedings of the National Academy of Sciences (PNAS)* would be published three days before my daughter Beatrice was born—and I continued to contribute anthropological data to other articles by Gordon Lab members, such as the *Nature* paper "Persistent Gut Microbiota Immaturity in Malnourished Bangladeshi Children." For the most part, lab members are identified by their real names throughout *Gut Anthro*—the Gordon Lab is too well known to try to anonymize it or the people who worked there. Most of the experiments that I discuss in this book became published papers that were prominent and widely disseminated, not only within the scientific world but in the popular press as well. Lab members have consented to use of their real names, except where noted.

I believed that investigating the biology and sociality shared between humans and microbes connected the concerns of anthropology and microbial ecology and enabled new modes of collaborative research across the natural and social sciences. I became interested in microbes because I was excited about how they could translate, how they mapped onto anthropological ways of asking questions of the world. I wasn't prepared for the ways in which microbial considerations would change the way I thought about and practiced anthropology. Joining the Gordon Lab became sort of "strong collaboration, that is, a form of collaboration in which explicit attention to the process is part of the project."[43] But, as anthropologists Leighton and Roberts write about their own work building and maintaining biosocial collaborations across public health:

> research collaborations between anthropologists and STEM scientists are, of course, not unusual.[44] However, in looking for examples

> prior to beginning our research, we found only a limited number of collaborations with engineering or public health that were led by cultural anthropology, rather than the other way around.[45]

The power hierarchy never tilts toward anthropology in these biosocial collaborations, precisely because all institutional structures, from funding to publication, empower the sciences. Doing ethnographic fieldwork among scientists, or "studying up," brings to bear a new set of political, ethical, economic, and disciplinary complications.[46] Special, privileged access allows for the cultivation of friendship, reciprocity, and collaboration, yet also increases the vulnerability of both sides: the ethnographer who may possibly be financially dependent on and certainly is ethically accountable to the scientist (as I was) and the scientist who has laid bare the practices and inner workings of science in progress (as Dr. Gordon did). Sociologist Des Fitzgerald and geographer Felicity Callard, in an attempt to move away from the (inter)disciplinary baggage of "collaboration," describe instead an "experimental entanglement" between neuroscientists and social science partners.[47] Like these authors, I found that the space one must occupy in a collaboration can be ethically, methodologically, and conceptually ambiguous.[48] My *experimental entanglement* with Dr. Gordon also was very rooted in Roberts's idea of bioethnography and what Emily Yates-Doerr describes as tactics of *careful equivocation*—attuning ourselves to disparate bio–social binaries but working together to shape global health imperatives.[49] These scholars have carefully constructed methodologies, trying to articulate what it is like to be in the field now as a medical anthropologist or STS ethnographer. There are no handbooks for this kind of collaborating. In my work with the Gordon Lab, anthropological stakes were continuously offset and commingled with concrete questions of health and illness, as well as highly technical metagenomic data. Anthropologist colleagues were impressed with my access to the lab and simultaneously viciously critical of how my relationship to the lab affected my research—as if they didn't have anything to do with each other. Everyone *wants* to collaborate, but no one knows how to actually *do* it.

Interdisciplinarity *CliffsNotes*

"You are not in the same boat." "Don't forget he has all the power and money." Before I left for the field, senior faculty and fellow anthropologists bombarded me with an abundance of various metaphors. Everyone except Dr. Gordon was very concerned about how I would fare in the lab. Scientific knowledge as an ethnographic object and anthropology's relationship to studying science now have a decades-long history of their own.[50] Since the 1980s, laboratories as ethnographic field sites and scientists as anthropological subjects/interlocutors have become a core of STS and a substantial subfield of sociocultural anthropology.[51]

Yet the relationship between anthropology and the scientists who anthropologists study remains understandably fraught. To the dismay of those forewarning me, there were power imbalances in my partnership from the start. Dr. Gordon was a very well-established, well-known scientist—a singular groundbreaker in his field (remember, father of the microbiome) with thirty years of experience. I was a humble PhD candidate, struggling to figure out how exactly to do fieldwork. The disparities were tremendous. How did the lab work fit in or conflict with my own research? I thought it was something I needed to figure out before I arrived in St. Louis. Instead, I spent the next fourteen months trying to answer these questions and a substantial amount of time trying to disentangle myself, before realizing that, much like commensal bacteria, those entanglements were part and parcel of the project itself. In this dynamic, I stand in for anthropology, and Dr. Gordon stands in for human microbial ecology, but how people relate on a personal, situational, subjective level makes or breaks the collaboration. Different personal, professional, and disciplinary arrangements have different outcomes.

Fortun and Fortun have discussed a "friendship with scientists"[52] as a mode for doing an anthropology of science that moves beyond a critical science studies concerned with systems of financial and epistemological domination. At the time, fresh from coursework and new to the field, I was drawn to their idea of friendship, and it was one of the frameworks through which I built my relationship with Dr. Gordon:

> To engage scientists on their own terms, as they work out what will count as true, rigorous, and worthy of concern. Friendship-conceived as a way of relating to others that demands reciprocity yet tolerates times out of joint, the not-always-predictable circuit of gifts, and the way exchange can work even when not a simple, reciprocal transfer that returns an investment offers a possible methodology.
>
> Friendship with informants is a rich tradition in anthropology, although often played out in contexts that call on ethnographers to assume the role of advocates for informants enduring tragedy and domination. Friendships with scientists within ethnographic projects are usually figured differently. Scientists generally have more cultural and legal authority, more resources, and more robust habits of collective action than ethnographers. Often of more importance, however, are the ways scientific knowledge is made, judged, and circulated, which can be both alienating and alluring to ethnographers. "Science" in public circulation often connotes certainty and transcendence of both history and culture. Science in daily practice is much more riveted by questions, doubt, and the necessary upsets of experimental techniques—especially, as today, in periods of rapid technological change.[53]

And indeed, the ways in which the Gordon Lab members practiced, talked, and thought about (but not how they published) science was incomplete, fractured, conditional. I started to see the importance of disciplinary *willingness*—the ability to put aside an intellectual ego and meet halfway, or sometimes three-quarters, or ninety-nine one-hundredths, of the way. It became clear that just by entering into our partnership, Dr. Gordon was demonstrating his wholehearted willingness. What I needed to give and what he needed to give were not the same. It feels necessary for me to mention Paul Rabinow and Gaymon Bennett's failed "experiment" in collaborating with the Synthetic Biology Engineering Research Center, within which Rabinow served as a CO-PI and received a substantial amount of funding over five years.[54] The anthropologists (quite acrimoniously) hold the scientists responsible for the failure of that partnership. When describing the project in 2009,[55] Rabinow also talks of willingness, but only in terms of questioning whether elite scientists were *willing* to develop a collaboration based on submitting themselves to "changes of a transformative sort in their habits and procedures."[56] Rabinow reads the power relations, upside down, and is then frustrated by the results. For him, the willingness of scientists is the key starting point; he doesn't ask anthropology or the anthropologist to undergo such a transformative change.

In their analysis of the evolution of their cross-disciplinary partnerships, anthropologists Leighton and Roberts work with the idea of trust as central to the process—the trust required from study subjects and interlocutors and being able to trust your collaborator's data and methods, even when they are very much not the same as yours. They also address frustrations and how to formulate expectations:

> We can't expect that a brief immersion in anthropological methods and theory will automatically lead our STEM colleagues to shift their epistemological moorings; nor should it. Working with other disciplines as equal collaborators meant starting from a position that assumes good intent on their part; trusting that their impulse to, for instance, produce knowledge that can be made into a deliverable intervention elsewhere in the world should not be immediately dismissed as "universalizing." And equally, it has meant opening ourselves to the possibility that they may challenge our approaches and assumptions in return.[57]

In my partnership with Dr. Gordon, as the social scientist, I was the one to do most of this maneuvering, as Callard and Fitzgerald describe in their chapter "Against Reciprocity: Dynamics of Power in Interdisciplinary Spaces": "Interdisciplinarity is entangled in much thicker structures of power than either its promoters or its practitioners are willing to recognize. Interdisciplinary collaborators need to give up not only the official fantasy of 'mutuality,' but also on that fantasy's self-consciously critical mirror image, *viz.* the idea that relations of power need either to be *overcome* or at least *faced up to* through reinvigorated forms of transparent dialogue, mutual respect, frank talking, and manifestations of emotions appropriate to the situation."[58] Instead these authors propose an intentional acquiescence on the part of social scientists, successful collaborations possible through facing the fact of disciplinary imbalances.

Within biosocial collaborations, mutual respect, and even care, is necessary, but more so an embrace of a destabilized methodology and uncertain outcome. Callard and Fitzgerald even suggest that instead of standing one's disciplinary ground, sometimes the most productive collaborative action one can take is adjusting and assenting to one's state of subjugation. For me, there were times I was swept away in the tidal wave of Dr. Gordon's dynamism, his excitement and expectations. It seemed we were truly going to transform the ways in which microbial science and anthropology were done together. We were going to change the world! Other times, when I sat in my office for days on end, meetings rescheduled, no way forward, I felt like one of many dependent children demanding the attention of a busy father. As a leader, Dr. Gordon was carefully transparent about his methods: "Transformative leaders set the tone for the group. I like to 'distribute enthusiasm,' not manipulate, but manage personalities." He valued what he considered goodness and personal maturity. When I asked him about his lab and staff management style, he responded:

> The evolution of science comes from a fundamental generosity, the sharing of ideas, and trust. Innovation happens through collaboration. It is important to empower who people are, not just what they do. Sometimes there is a lack of clarity about how things will actually work out, but I don't have a top-down vision of science—no thunderbolts. Members of the lab set the tone, and need to have ownership to promote motivation. Leadership is about giving. This is how it works in other creative fields like art and music.[59]

Dr. Gordon is sincerely a good person; he encourages, nourishes, and supports people. He gives a lot and expects a lot in return. He is everything you would want a person in a position of power to be: respectful, communicative, kind. Even so, there will always be something unbalancing about this sort of arrangement. Everyone in the lab was a planet that orbited a life- and career-giving sun. Collaboration became a matter of deciding for yourself if you could be one of those planets—if you wanted to work this way. Callard and Fitzgerald incisively (and brutally) discuss this state of being in their descriptions of collaborating with neuroscientists: "if you are going to live in such spaces, *better to learn to live in them as they are,* and give up on an agentic fantasy in which you will be able substantively to transform imbalances, inequalities, and existing norms governing epistemic (and other kinds of) potency."[60] I'm not sure if these are inevitable collaborative realities. Maybe something else is possible? I didn't work with Dr. Gordon because I thought anthropology was finally going to get a seat at the biological science table. I don't think I ever had delusions there was parity, between our disciplines or between us as individuals. The fact that I wasn't a scientist made the relationship I had with Dr. Gordon unlike any other in the lab. I was Pluto, orbiting on the very edge of the solar system, my very status as a planet under interrogation. But Dr. Gordon never made me feel as if I were at the outer limits; he cared deeply about the anthropological work, and about me. And still, my project was surely the last priority on a long list of things to do. When grant applications (which impacted all of us in the lab) were due, when scientific papers with other lab members were being prepared for publication, when experiments were in full swing, the anthropology (understandably) slipped to the bottom rung. My point here is not to qualify my work with the Gordon Lab (or collaboration in general) as good or bad but to describe the acute complexity of practices involved in such a relationship.

I am and always will be grateful to Dr. Gordon. And that gratitude will always be tangled up in frustration, misunderstandings, and struggles for independence. Dr. Gordon's unilateral control over every aspect of his lab has contributed to his success; he is a savvy fundraiser and a judicious manager. From funding to technology to the hiring of lab members to breeding the mice and making the mouse food—he made sure that every element of the lab was constructed to exacting specifications. Even the layout of the physical space in the Center for Genome Sciences was carefully crafted. His office had glass walls and was located at the entrance of the lab—he could see who came and went, and his lab members could always see him. The shared conference room was also all glass and referred to as the fishbowl because of the visibility of those meeting inside. The lab itself was open, with benches on casters, and Dr. Gordon liked to equate the space with the kind of work they did, describing "moveable labs, flexible science." The eating and gathering places, such as the lunchroom and coffee lounge, were strategically placed to encourage cooperative working-while-socializing. He placed certain lab members at certain benches and assigned particular rotation students to specific postdocs to facilitate partnerships he felt would be fruitful. One person told me, "People think they have choices about where they sit—but they don't." This wasn't an antagonistic commentary but an observation about the intentionality of Dr. Gordon's influence.

I did not enter into our collaboration unawares. There were moments when I was faced with the reality of our power imbalance, moments of anger, discouragement, and defeat. At times, I felt collaboration was an impossible endeavor, but these comments also show my inexperience and impatience with disciplinary parameters, with what I thought I was sacrificing. I didn't realize at the time how much I was getting:

> I've started to really feel that this whole collaboration is about anthropology becoming more science-like. And what is science giving? Or giving up? They might begrudgingly include an aside about anthropological information in a paper, but not in a study design, not in their data. I just got a sense that what I *really* want to do, the work and writing *I* care about, is always going to take the back burner.
>
> The thing is, before I came here, I strained against the limits of anthropology. I never fit in, and had a lot of problems with the way things are done in the discipline. But now I feel like I have to defend it! Because I don't want to be a scientist either—there's so much bullshit I don't agree with at all. So where does that leave me?

I began to participate, uncomfortably, in what I came to believe was important science being done, to bring anthropological perspectives to bear on this science, and make both disciplines better through the process, in what we came to think of as an anthropology of microbes. And Dr. Gordon's goals? I believe they were to employ and learn from someone with expertise in human sciences—to learn how to think about the people affected by and affecting his experiments. We tried to locate an emergent problem space shared by our disciplines, and we agreed to be accountable to ethical and effective scientific practice. As Nowotny, Scott, and Gibbons argue in *The New Production of Knowledge* and *Re-thinking Science,* science in the mid-twentieth century shifted modes, from "mode 1," knowledge-driven, fundamental research, to "mode 2," context driven, interdisciplinary, and concerned with applications of research in the real world.[61] The Gordon Lab not only epitomized this mode shift; Dr. Gordon pushed far beyond the boundaries of problem-motivated interdisciplinarity. At the 2010 BioMed Symposium at Washington University, Dr. Gordon discussed his vision for individualized diagnoses and therapies, citing the interdependency of disciplines and new social structures required to do such science. His unprecedented biomedical agreement with the pharmaceutical company Pfizer (and then, later, food companies like Mars) brought academia into an exciting but murky relationship with industry. Dr. Gordon saw the biomedical work at Wash U as a seed for a national initiative: "The NIH has a vision of a network of clinical and translational medicine centers. They are begging us to do these sorts of things." A contestable term,[62] *translational medicine* has been described as biomedical and public health research that aims to improve the health of individuals and the community by "translating" findings into procedures, policies, and education; the rapid transformation of biological knowledge into effective health measures through collaborative, multidisciplinary approaches; and, most simply, bench-to-bedside science.[63] Translational medicine has become a priority for the NIH, which established the National Center for Advancing Translational Sciences (NCATS) in 2012, funding more than sixty centers with $500 million. NCATS calls translational medicine the faster delivery of new treatments and cures for disease to patients.

Overall, Dr. Gordon's focus on global human health, the sustainability of our populations, and what we will eat demonstrates his "mode 2" research drive. As one grad student told me:

> Jeff's philosophy of science is, always be ahead of the curve. Jeff has different impulses from reductive thinking—but humanization is very complicated. This is a hypothesis-generating lab rather than a hypothesis-testing lab. What intimidates about this ethic? The expense, the focus on computing, math, data sets. New knowledge and interdisciplinarity is needed. Like micro arrays—they are so expensive for something, and you're not sure what the outcome will be. What's most different about CGS [Center for Genome Sciences] is that most big sequencing centers control distribution and data production, but Jeff is big on democratization, which has created tension with the HMP [Human Microbiome Project]. Democratization necessitates knowing a bunch of new science, for example, computer science. Biology without hypothesis testing? It's fuzzy. But no one can refute your results because no one else really has the knowledge.

Ethical Positioning

As discussed previously, I worked for several months in Dhaka, conducting my own ethnographic research, as well as serving as the Gordon Lab anthropologist. My employment at the Gordon Lab placed anthropology in the scientific pipeline, as a link between information about Bangladeshi food choices, changes, and traditions and genome scientists and epidemiologists. My lab role had dual parts: first to provide qualitative ethnographic data about living conditions in Mirpur and second to help Dr. Gordon think through more ontological, ethical aspects of the scientific work. As Fortun and Fortun describe:

> it is at the level of scientific practice that ethnographers can tap into the ethical dilemmas and imaginaries of the sciences, helping draw out critical articulations made by scientists themselves,

> helping scientists think through the ever-important tension between stabilization and change in the sciences, and providing discursive resources for engaging the ethical, legal, and social implications (ELSI) of science.[64]

Dr. Gordon was interested in how the invisible world of microbes was made manifest through what and how people eat, finding out what resources are being used to find out what people are eating: databases, contexts, variation. He wanted to know how families were defined. His anthropological interest shifted topics over the course of the year; at times he was engrossed with the idea of how invisible microbial worlds would be understood by Bangladeshi mothers, whereas at other times he was interested in the temporality of culture. I wanted to explore everyday life in Dhaka, asking how people were affected by their enrollment in the microbiome study and what they did with the health and nutrition information with which they were inculcated—if they incorporated, rejected, or hybridized it. My research was much more abstract and theoretical than the lab's. I was trying to find inroads to my bigger questions about the scaling of human–microbe relationships, in a place where microbes were both conceptually nonexistent and a practical matter of life and death.

By the time I finally arrived in Dhaka for the second time, I had been vetted by Dr. Gordon, approved by the New School's IRB, trained and authorized by Washington University's Human Research Protection Office, and thoroughly scrutinized by icddr,b's Research Review and Ethical Review Committees. I had to complete human subjects research training and two separate lengthy IRBs for the two different sets of subjects: Gordon Lab scientists and Mirpur mothers. A social scientist embedded in a biomedical study can create uneasy circumstances for all involved. The IRBs understood, in the context of the Gordon Lab work, the taking of blood, urine, and stool. The ethical risks and protections seemed uncomplicated and clear. But they were stymied by how to safeguard the rights of subjects if what was to be taken were opinions—especially when these people were also the subjects of scientific research and their participation in that very research was something I would be asking them about. Anthropologists have long chafed against what are seen as bureaucratic and ill-fitting IRB requirements,[65] and understandably; several months of my fieldwork focused on various human subject ethical reviews. When I arrived in Dhaka, the absurdity of the IRB and ideas of informed consent became clear; some mothers in the study couldn't read; community structures and communal living meant that even if the homeowner consented to my questions, at least three or four or ten neighbors would stick heads in windows and chime in to the conversation. The larger, looming concern with ethical practices revolves around why anthropologists become the experts on ethics in scientific collaborations—a mantle we sometimes involuntarily shoulder, and other times one we presumptuously snatch. In a *Cultural Anthropology* series on "Collaborative Analytics," Fiona McDonald and coauthors discuss an anthropologist's responsibility in collaboration:

> No two collaborations are ever the same. Scales of responsibility are situational and subjective. Therefore, we feel that thinking about responsibility needs to happen on a case-by-case basis. We all know about and uphold the embedded ethics of our work as ethnographers to protect participants and collaborators, we adhere to ensuring that realistic expectations are set and met when working with new communities, and we conduct research as objectively as possible. This is not the responsibility we are speaking of here per se. What we mean is, let's each ask ourselves when we collaborate: To whom are we responsible? Who is responsible to us? For how long are we responsible for the new form we are engaging in? Do we have a responsibility to the knowledge we are producing? And what is our duty of care for the audiences we are working with?[66]

As my extensive IRB requirements show, my responsibilities were manifold—to the mothers and researchers in Dhaka and, simultaneously, to Dr. Gordon and the Gordon Lab members. These obligations were not mutually exclusive and in fact sometimes uncomfortably contradicted each other. My position (both physical and professional) in the lab was also awkward. I had an office catty-corner to Dr. Gordon's, where I could see and hear everyone's comings and goings. Lab members were unsure if they could refuse to talk to me, if they had to be nice, or how much they should reveal in formal interviews or casual conversation. They didn't know what Dr. Gordon wanted them to say or not say to me. Operating with anthropology-101 knowledge of my discipline (or none at all), most scientists didn't understand what I was doing there or what I expected to learn. One postdoc later told me, "At first I thought you'd be hiding out in a nitrogen tank in the corner taking secret notes or something." The scientists were curious, and all asked if I'd been to other labs; they wanted to know how they compared from an anthropological purview. People in the Gordon Lab often staunchly reaffirmed that I was a "cultural" anthropologist (read, not a *real* scientist as perhaps a biological anthropologist would be).

Lab members weren't the only ones with questions; I also was confused about my position, my access, and sometimes my purpose. At times I was colleagues with these scientists, sometimes a coworker, and always an observer. When positioned by Dr. Gordon to work or write with other lab members, I was met with curiosity, and often with confusion. The graduate student analyzing the Bangladesh samples was extremely helpful and forthcoming when I asked about his project, yet before Dr. Gordon's suggestion that we work together, he never even briefly considered incorporating any of my findings into his paper. This was a matter, not of willingness, but of disciplinary data disjunction—he just couldn't envision how ethnographic information would be useful to him. After many conversations and explanations during my year in the lab, it is possible that those working on projects related to malnutrition or human studies did begin to see the anthropological value to the work they did. But I also learned that Dr. Gordon was unique; no one else in the lab cared as much about a social science contribution. Unexpectedly, junior scientists early in their careers were much more likely to be rigid in their disciplinary alignments—perhaps because they felt like they had more to lose.

Intra-laboratory-action

Washington University was institutionally interested in interdisciplinarity, especially within the sciences. Collaboration was a hallmark of the school, with mottos such as "Excellence in science, service to others." One graduate student explained it to me this way: "There is an institutional culture of openness at Wash U—people are not interested in competing; departments don't do things like hire five people for one tenure-track position. No one knows about St. Louis, so everyone has to work together to build a reputation."

This university-wide culture, combined with the ways in which metagenomic studies of the microbiome require a cofunctioning of different kinds of scientists, and Dr. Gordon's own collaborative ethic, made the Gordon Lab an environment rife with cooperation. An immunologist in the lab commented, "In other labs, people work on the same things, become competitive and begin to think about things in a similar way. Not here. Everyone comes at things so differently, such different backgrounds and disciplines." At its best, this kind of work allows thinkers across fields to contribute to a common goal. Ian Hacking describes this sort of thing in his response to a rising tide of collaborative scholarly work and the population of ideas of interdisciplinarity:

> In my opinion what matters is that honest and diligent thinkers and activists respect each other's learned skills and innate talents. Who else to go to but someone who knows more than you do, or can do something better than you can? Not because you are inexpert in your domain, but because you need help from another one. I never seek help from an "interdisciplinary" person, but only from a "disciplined" one.[67]

In the beginning, Dr. Gordon sought out my expertise because he was interested in social science work. His thoughts about lab members' capacity to do this "social thinking" changed over time. Dr. Gordon was a strong proponent of the democratization of technoscience and believed strongly in leading by encouragement. Lab members had large budgets to work with and access to genomic sequencing technologies. However, obtaining such extensive funding required careful negotiations and prickly financial collaborations, between the Gordon Lab and nonprofits like the Gates Foundation, big pharma companies like Pfizer, and major food manufacturers such as Kraft and Mars. Academic science's need for new biotechnologies, monies for capital equipment, and access to global populations can show how funding encourages, forces, and fabricates collaborations.[68] Nonprofit funders often became impatient with mouse models, finding the results not applicable enough to human health and pushing for human subjects, but as one postdoc commented, "human testing is so much harder—the results are less reliable because of so many uncontrollable variables, and the extreme IRB complications reduce the value of human studies." Conversely, funders like major food manufacturers were very invested in mouse studies. In fact, a major benefit to the lab of partnering with "industry people" was that these companies had the technology to pellet selected human diets into mouse chow. However, tension with these funders emerged around agreeing on diet ingredients for experiments that would have meaningful outcomes both for scientists studying the microbiome and for corporations trying to figure out how to increase market share. By 2019, the Gordon Lab published a paper in *Cell Host and Microbe* that drew conclusions about how specific gut microbes could respond to and degrade components of processed food that had potentially harmful effects on human health.[69] The paper ends with a suggestion that perhaps microbes could be prescribed and function as probiotic counterbalances to the dangers of processed foods. Framing gut microbiota as tools by which elements of *bad* or unhealthy food can be transformed into "innocuous metabolic products"[70] sidesteps all the important histories about the intentional distribution, political economy, and ubiquity of processed foods in poor communities and communities of color. It ignores the globalization of processed foods and the replacement of indigenous or local ingredients with prepackaged, "westernized" food. In these ways, "collaborating" with industry becomes a powerfully precarious way of getting science done—a large part of the mandate for translational medicine. But it also constrains what trajectories the science can take, for instance, cultivating and ingesting specific microbes to neutralize processed food, instead of exploring how to make healthier food more accessible or even reconsidering what health means in different places.[71]

Science in this format—in a well-funded, prominent, innovational lab—was at times very creative. Because so little was scientifically known about commensal microbes (and what *was* known was known because Gordon Lab members or their collaborators figured it out), there was a lot of latitude. In 2010, those working in the field of human microbial ecology tended to converge, and the Gordon Lab has several collaborators with whom they worked on most of their experiments. Some of these partners included icddr,b; the Knight Lab at the University of Colorado (which did all the bioinformatics work for the Gordon Lab); and other scientists in nutrition science, genetics, and so on at the University of Virginia and the University of California, Davis. At a cross-institution site visit, a scientist from a collaborating lab told me that other labs are not "interdisciplinary" in the same way that the Gordon Lab is. He told me the Gordon Lab has a method of implementing new technology early on and exploring the limits and usefulness of it. "They can really let their imaginations go in terms of 'what would be cool to know.' Things are changing all the time there, not like some microbiology labs where everyone is working on the same protein for years—there every postdoc has his or her own unique project."

Running a lab in this way gives junior scientists and graduate students expansive and unique access to funds and technology and fosters an interdisciplinary group, bringing microbiologists together with ecologists, geneticists, immunologists, and gastrointestinal physicians. However, these technologies are also necessarily limiting, focusing on bacterial genomes and determining the function of those bacterial genes in human biological systems. These methods, kinds of data, and scientific discourse were unfamiliar to some lab members, who were brought into the lab to contribute their different disciplinary viewpoints. Some scientists came into the lab with no experience with mouse models, others with no background in DNA extraction or sequencing.

"It's hard to deprive animals of micronutrients. They store iron in their livers, they know how to survive. On the other hand, the Malawi diet is so low in protein, zinc, vitamin A, and micronutrients in comparison to RUTF [ready-to-use therapeutic food] or standard rodent diet, that when we first put mice on it, they had no fat. They had no pregnancies. I had to make a breeder diet, adding soybean oil, thiamine, and the amino acids that are deficient. These are the only mice on the Malawi diet that have been able to breed so far, and the pups are stunted. None of the pups have been able to breed." Michelle Smith is filling me in on the state of her experiment as she and I stand next to an industrial-sized food mixer called "Chowbacca," in a special room of the Gordon Lab where mouse diets are made. Michelle is a third-year postdoc and, when I was in the Gordon Lab, had the only highly humanized study in the lab, switching the mice between the Malawi and RUTF diets. In the years since I left the Gordon Lab, most projects in the lab have become focused on humanized mice (and gnobiotic piglets). Specifically, many experiments emerged that gavaged mice with the fecal samples from the Dhaka birth cohort, oscillating those mice on "Bangladeshi" and RUTF diets, and comparing results from the healthy and malnourished donors. *Gavage* is a term used ubiquitously in the lab; it means to force-feed food or drugs, particularly to animals, and especially with an implement down the throat. Michelle is interested in micronutrient deficiencies, individual ones like iron absorption.[72] Michelle is looking at mouse urine, leptin in blood, and ceca to see the turnover of microbial populations. The cecum is the pouch connecting the end of the small intestine (ileum) with the beginning of the large intestine (ascending colon). She is finding that nutritional deficiencies change the physical structure of the gut; in a "healthy" intestine, the epithelial cells have a large surface area, facilitating greater absorption of nutrients (epithelium is the thin tissue lining the alimentary canal and other hollow structures of the internal body). The villi in intestines with malnourished epithelial cells were blunted, smaller, and shorter, providing less surface area and therefore less nutrient absorption. Michelle wanted to find out if the microbial communities were also affected by this structural change.

When I ask her how what she's doing in the lab differs from her graduate work, Michelle tells me, "Careers are made on postdocs, not PhDs." Her background is in molecular biology (her dissertation was about the regulation of apoptosis in postmitotic cells), but here she has learned all about big data and sequencing analysis. When Michelle has her own lab, she'll get back to her original work—right now, she has access to the Malawi samples, access to sequencers, and her experiments cost hundreds of thousands of dollars, funds she likely won't have in her own lab. She anticipates that a whole grant in her own lab would be as much as one Gordon Lab experiment. "When I have my own lab, I'll be able to go back to my other work, cheaper and slower, take more time." She has a clear understanding that the resources in this lab are remarkable and that the rest of her career may not be so flush.

Michelle told me, "I really don't like computational biology, I mean really, really don't like it, and that was a major component of the work we do more so than I had really anticipated. The tools we have now to study the microbiome are very dependent on large-scale 'omic data sets, and that means a lot of data crunching." Though she had no interest in big data science, she quickly adapted to the fact that all Gordon Lab members must learn some bioinformatics. I asked four other lab members about what unexpected types of work they found themselves doing, and they said the following:

> Almost everything I do and the work that I've done in the Gordon Lab has been a departure from what I was "intending" to do when I came to graduate school at Washington University. I am now working with bacteria. I am studying commensal relationships. I have learned a lot more about metabolic and nutrition sciences than I ever thought I would.

Gordon Lab members call the industrial-sized mixer used for making different diets for laboratory animals "Chowbacca."

Since I first worked with microbiology, I knew I was interested in microbes, and I have always had an interest in human metabolism. However, I never thought I would combine these two passions through poop and computational biology.

I'm trained as a field biologist, which is just an entirely different profession. It's a lot messier, and based on real-world things, not so much on following protocols exactly and measuring precise volumes of liquids. My PhD work was done sitting in a Missouri swamp. There is a difference in working on bacteria instead of plants—it requires cleanliness because bacteria are everywhere and can mess up your DNA. My background is in evolutionary ecology. There's never a one-to-one correlation between the science and the goals.

These comments show how complicated scientific interdisciplinarity can be. Though these scientists may ask different questions of the data, and despite the objective of harmonious collaboration, ultimately, everyone must use the same technology and do the same sorts of experiments involving germ-free mice and metagenomic sequencing. Every postdoc and graduate student member of the lab was able to develop their own experiment—but within parameters. In 2010–11, ten postdoctoral fellows and ten graduate students were working on a variety of gut microbe experiments. An experimental theme was dictated by resources, technology, and the overall lab ethic/doctrine—to develop new gut microbiome–directed therapeutics that improve human health. This was not always comfortable or productive for all lab members. One trade-off for being cutting-edge was operating in a contextual void, as one frustrated lab member told me: "Jeff works bottom up, everyone else in the field works top down—complicated diets, complicated microbiotas—no one else is doing it. Science relies on seeing how other people fail, and this lab doesn't get the benefit of that."

Global Collaborative Frictions

In Dhaka, nineteen FRAs worked full-time with the mothers in the Microbiome Discovery Project in Mirpur. The ways in which the study is explained to potential subjects has very little to do with the ways in which the Gordon Lab or icddr,b scientists describe the research. I asked all the FRAs what they tell the mothers about the study when recruiting them. Zarat said, "We explain that their children are not well, that they are malnourished. Children have no appetite, won't eat, aren't healthy. We explain that we will provide *pushti* packets and find the causes of malnutrition in blood, urine, and stool samples. Sometimes there are genetic factors for malnutrition, we explain this to every mother." When I asked what they tell mothers about the microbiome, Roji said, "They know a little bit. They know that there is bacteria in stool. We explain that to them, explain to them about handwashing, and that hygiene is important to prevent the spread of bacteria, and that if they don't, they will get pain in their stomach." Sunita, who has a degree in accounting from National University and worked at the BRAC Child Nutrition Program before coming to icddr,b said she tells them about cleaning, washing, and hygiene, but she doesn't explain about bacteria: "I tell them that not being clean, not washing their babies can make them sick."

This brings to mind an important question Alex Nading asks in his article "Evidentiary Symbiosis: On Paraethnography in Human–Microbe Relations": "How might one do a social study of the microbiome in places where it does not (yet) exist as a category of expert practice or public discourse? Strictly speaking, the microbiome, as a category of scientific and public interest, has been limited to the Global North."[73] It is precisely the critical ethnographic contribution to the microbiome project in Mirpur that aimed at addressing this disparity. Taking seriously what Nading calls *evidentiary symbiosis,* my development of an anthropology of microbes became a practice, "in which the cultural/interpretive evidence of paraethnography interfaces symbiotically with the quantitative/statistical evidence of bioscience."[74] For Nading, *paraethnography* is all the social science knowledge creation that doesn't fit organically or logically into the scientific models. Merging the idea of paraethnography with Roberts's bioethnography, I continue to devise ways in which my ethnographic data can correspond with, disrupt, and transact with microbiome science—in the form of scientific paper appendices, conversations, redesigned community interventions, and so forth.

In Mirpur, FRAs taught the mothers the word *jibanu,* in Bangla literally "causative organisms." *Jibanu* is not a word typically used in Dhaka outside a medical setting and is not exactly the same as *bacteria, microbes,* or *germs.* When the FRAs recruit mothers and mothers-to-be, the microbiome study is not explained in scientific terms, as FRA Roji explained to me:

> I give them information about the support they will receive in this study, what problems we are addressing—malnutrition—what we will sample from the baby, etc. If they agree, they will sign the consent form. I explain that their children, for case control, are nutritionally very poor, that they will need to come to the Pushti

> Center, and I identify if they need to get supplements. Every month we measure the stool to see if there is any problem, and see what they have been eating. The mothers think that having the stool taken is very good for the child's health. They think they can evaluate the child's health through the stool sample.

Pushti Centers are the three Mal-ED nutrition centers distributed throughout the five Mirpur camps. Mothers of case-control infants bring their children to these small centers (green paint, water filters, food, nutrition, and health care educational posters) to receive *pushti* packets. *Pushti* packets are locally produced RUTF macronutrient meals: 150 kilocalories consisting of rice powder, lentil powder, molasses, and soybean oil.[75] They also receive Moni-Mix micronutrient "sprinkles" to add to home-cooked meals.

It is within the network of FRAs and mothers that the Mal-ED project is both carried out and subverted. The dynamic, collaborative interplay between these women is on the surface class-dominated, but it is also entangled with the scientific project, the bodies of mothers and babies, and a bigger picture of ubiquitous NGO intervention in Bangladesh.

As highly educated employees of icddr,b, the FRAs do not live in Mirpur but in more affluent sections of Dhaka like Gulshan and Banani. However, icddr,b relies on a commonality of Bengali womanhood, on the duties of motherhood, to foster a successful study relationship between the mothers and the FRAs. Ninety percent of the population of Bangladesh are Bengali Muslim, as were all the study subjects in the Microbiome Discovery Project. Bengali is the predominant ethnic group of Bangladesh, but there are also religious minorities, including Bengali Buddhists, Bengali Hindus, and Bengali Christians. Non-Bengali Muslims and Indigenous ethnic groups make up the rest of the Bangladeshi population. Among the Indigenous (or Adivasi) populations are the Chakma, Marma, Santal, Mandi, Tripura, Mro, and Bisnupriya Manipuri. Besides the Tripura, who practice Hinduism, most Adivasis in Bangladesh are Buddhists who follow the Theravada school of Buddhism. Indigenous and minority populations live primarily in the rural north, south, and Chittagong

Hill Tract areas of Bangladesh, and I was working only in Dhaka City.[76] It is important to distinguish between "Bangladeshi" and "Bengali," as they are not equivalent.

Like Maliha, her FRA Zarat also prepares dinners with a cast-iron *boti* knife on the floor of her kitchen, she also has a daughter whom she marked with ash at birth to protect her from the ill will and misfortune of the evil eye. There is a compassionate proximity between these women, a bodily closeness as they sit on beds, talking. While samples of blood and stool travel in only one direction, information is currency between these women. The FRAs look to these mothers for answers, as Adeela, the FRA supervisor, tells me: "Mothers are the best people to know what kind of illness their child has, and what kind of treatment they require. I have learned a lot from working on this study." The FRAs are empowered by the authority of icddr,b (and, by extension, of the Gordon Lab and the Gates Foundation), and by the science itself, but they also know that the study mothers have valuable knowledge and expertise to share. This knowledge finds its way into the data through the FRAs. The dominion of icddr,b, the responsibility to American research institutions, and the unequal distribution of labor also weigh heavy on field researchers, as Adeela complains: "There are 600 enrolled in case control, and 175 in cohort. We have nineteen FRAs. Our supervisor says that is plenty! They [FRAs] do so much work, so many types of questionnaires, so much paperwork to do. This is important research; we have to be very careful to not make any mistakes."

How the FRAs understand (and practice) the objectives of the study is not exactly the same as how the directors at icddr,b understand them, nor are either of these the same as Gordon Lab scientists think about the research goals. And all these differ from the aims of the program officers and, ultimately, the founders at the Bill and Melinda Gates Foundation, which funds Mal-ED, the Microbiome Discovery Project, and much of the Gordon Lab work. Those "impatient optimists" (the Bill Gates self-proclaimed driving attitude behind the foundation)[77] are primarily concerned with the intersection of malnutrition with enteric diseases and the impact of malnutrition on vaccine efficacy and physical and cognitive stunting. The Gates Foundation program officers sometimes conflict with the scientists on the project; they argue about mice and methods and dispute the value of animal models in rapid translational medicine. Sociologist Carrie Friese writes about this quandary in her article "Realizing Potential in Translational Medicine," asking, "Why should we support the use of animals in preclinical research if this program of research is not resulting in improved clinical care for humans?"[78]

As the primary funder for projects like Mal-ED, the Gates Foundation has the power to "speed the translation of scientific discovery into implementable solutions"[79] by developing and implementing alternate systems for processes, such as clinical trials. The reworking of these practices is both productive and troubling. Unsubtly, and through powerful financial means, the Gates Foundation directs scientific research toward areas where it can have the most impact and to "accelerate the translation of discoveries into solutions that improve people's health and save lives."[80] Paradoxically, though the Gates Foundation boasts extensive "cultural competency" (training for health professionals to interact with culturally diverse patients) and even "structural competency" (health professional education that focuses on systemic inequities and "upstream" social determinants of health), programs like Mal-ED lack what Yates-Doerr calls "translational competency." Translational competency asks, What is health? Who has it? and How does health emerge and transform in practice?[81] Yates-Doerr advises, "Gaining translational competence involves joining concern for cultural difference with concern for structural inequity, and understanding how medical structures, which are tied to histories of violence, colonialization, and dispossession, will value some cultures over others."[82] Gates Foundation translational medicine prioritized scalability and universality of treatment over this sort of nuanced reflexivity. At a Gates Foundation site visit, I heard a program officer say they would ideally like to see something like a "malnutrition vaccine," literally, a comprehensive, easily deliverable malnutrition solution. Contrary to this, those at Washington University, the University of Virginia, and the other scientific research institutions composing the Mal-ED consortium are invested in investigating the genetic, environmental, and microbial contributors to malnutrition. They are thinking about the biological pathways involved in malnutrition and how to use metagenomics, mouse models, and genome-wide association studies. This is one example of collaborative tensions between academic scientific researchers and public health NGOs.

The doctors and nutrition experts at icddr,b have a stake in the science but an equal share in the overall health of their constituents and in improving the quality of life of those in the communities they serve. icddr,b is tackling the larger issues of countrywide poverty, malnutrition, and lack of health care; its overall mission is to "harness the power of high-quality research to address the health problems of Bangladesh."[83] The Cholera Research Laboratory in Dhaka (which became icddr,b in 1978) developed oral rehydration solution, or ORS, for the nonintravenous treatment of cholera and other diarrheal illnesses. ORS is a simple saline and sugar solution that can be used at home to prevent potentially fatal dehydration during episodes of enteric disease—revolutionary for countries where access to medical care is poor and incidents of diarrhea can be persistent and life-threatening. Packets of ORS are commonly kept in households in India and Bangladesh. It is estimated that ORS has saved tens of millions of lives and reduced the risk of death from diarrhea by 93 percent.[84] In her analysis on biopolitical governance in Bangladesh, historian Michelle Murphy has asserted that certain biomedical/technological developments (she uses icddr,b's invention of ORS as her specific example) move "prevention of death" logics from infrastructural problems, such as water management, to the disbursement of emergency medicine. She employs the term *appropriate technologies* to criticize tools like ORS for their elision of any further examination of problematic infrastructures in low-resource settings.[85] The concept of appropriate technologies in development studies was created to describe low-tech, cost-efficient solutions designed with the cultural, socioeconomic, and environmental specificities of the setting in mind. However, Murphy and others caution that provisional and inadequate stopgap solutions sometimes rigidify into appropriate technologies, thus making deficient, temporary fixes permanent.

Clearly the material practices of invention, care, and disease treatment deserve a careful consideration. Murphy's move to point out the danger of shifting contingencies and intervention priorities is an important one, and it is equally important not to footnote the millions of deaths prevented through the use of ORS. ORS has been described in *The Lancet* as "potentially the most important medical advance of the 20th century"[86] and is estimated to have saved more than fifty million lives worldwide since the 1960s. Certainly a medical anthropological analysis needs to examine claims like these. Bangladeshis in Mirpur should not be defined by their poverty, and Dhaka should not be defined by its population density and disease burden. But real, bodily conditions must be contended with for survival: "The adverse outcomes associated with structural violence—death, injury, illness, subjugation, stigmatization, and even psychological terror—come to have their 'final common pathway' in the material."[87] This debate shows the complicated contradictions of biosocial collaborations, as well as the friction of working in the spaces of real life and theory. Anthropologist/physician Paul Farmer posits, "there is little compelling evidence that we must make such either/or choices: distal and proximal interventions are complementary, not competing. International public health is rife with false debates along precisely these lines, and the list of impossible choices facing those who work among the destitute sick seems endless."[88] Social scientists must look critically at the humanity-saving declarations of biomedicine, but we must also examine the representational violence in which we as scholars continue to participate to make careers for ourselves. Theory enables convenient disconnections from the very real violence within which our interlocutors sometimes live and the ways we gladly trade on their life-and-death circumstances to build tenure portfolios.

In cases like Bangladesh, appropriate technologies like oral rehydration therapy are necessary, sometimes perpetual and, yes, insufficient steps. They can also be seen as complementary to, rather than competing with, large-scale infrastructural change, "working at multiple levels, from 'distal' intervention—performed late in the process, when patients are already sick—to 'proximal' intervention—trying to prevent illness through efforts such as vaccination of improved water and housing quality."[89] As an organization run by Bangladeshis for Bangladeshis (but funded internationally), much of the work icddr,b aligns with this; their institutional strategic plan for 2019–22 included goals like "develop and promote use of innovations for Bangladesh and the Global South," "increase the visibility and impact of our research evidence," and "invest in our people." If, as so many scholars have been pointing to, the biological and social are unavoidably coconstitutive and people, things, and agency come into being through their interactions, can social scientists make moves toward ethical collaborations with scientists where biomedical and socioeconomic interventions are cooperative and coprioritized? Investigating all the intra-acting parts of the microbiome can utilize Farmer's idea of an anthropology of structural violence to figure out real-world and scientific interventions that matter to people, treat them ethically and equitably, and, as Yates-Doerr proposes, make a "long-term commitment to people and problems as they change over time."[90]

My participation in the malnutrition microbiome study in the Gordon Lab, coordinating ethnographic information with big data, was conceived as a collaborative move toward proximal interventions. This involved accounting for how cultural practices like cohabitation, nursing, cooking, and feeding might show how gut microbiota are amalgamations of everyday life that shape the evolutionary cohistories and genomes of humans and microbes. Ethnographic methods forced me to consider whether it is it possible to keep the underlying causes of malnutrition—poverty, household food insecurity, inadequate health care, sanitation—at the forefront while studying the long-term consequences of diarrheal infections on gut microbiota, human genetic polymorphisms which might affect nutrient absorption, and links between intestinal microbial ecology and childhood undernutrition.[91] Anthropologist Cal Biruk faces these shifting ethical obligations, especially in the context of collaborations with scientific researchers:

> Our commitment to critique and political alignment with the disenfranchised should not exculpate us from engagements and transactions in the field that activate—rather than redress—historical memory and expectations. For survey participants and anthropological informants alike, ethics are merely a starting point for difficult conversations that should aim to center historically informed accountability.[92]

This is such wise advice, to take ethics as a starting point. And so my partnership with the Gordon Lab required detailing discrete social, political, and environmental conditions to highlight changes in both those conditions and the bodies embedded in them. I failed, a lot. Sometimes this research worked against my idea of an anthropology of microbes, as did endless requests for quantified data. Once, after the hundredth such request, in frustration, I blurted out, "OK, well, eight out of nine families I talked to used toilet paper. Is that valuable quantitative data?"

The FRAS were primarily concerned with babies. When asked about the reason for the study or what they thought the results would be, they unequivocally answered, "To improve the health of children." This is not entirely different from the Gates Foundation, icddr,b, or Gordon Lab. But it is not entirely the same. *Malnutrition,* then, could be seen as the obligatory passage point,[93] a point of necessary agreement where the interests of many actors (in this case, mothers, FRAS, icddr,b administrators, Gordon Lab scientists, and Gates Foundation officers) converge into one workable idea from which the network can be formed and action taken. All actors are focused on addressing "malnutrition" in some form, but as discussed later, the FRAS operationalize independent methods and goals. Further complicating the network is the fact that the FRAS and mothers are those implementing the study practices, deciding what is important, and collecting the data. While the samples originate with the babies (and, in some ways, with the mothers through their breast milk), it is the mothers who are responsible for reporting about their families' food consumption, illnesses, and daily lives. The mothers scrape stool into plastic vials in their crowded, hard-to-keep-clean houses with no doors and walls made of old newspapers. As has been discussed elsewhere,[94] it is mothers who, through eating and feeding, are accountable for the health of their children and the biological outcomes of generations.[95]

Mirpur mothers were (for the most part) gratefully enrolled in a project in which they received free health care, medicines, and an almost guaranteed better health outcome for their children. It is also possible that they emphasized their gratitude in their interviews with me, for fear of exclusion from the study. Unanimously, all the mothers with whom I spoke stated that access to free medical services was the primary motivation for enrollment in the Mal-ED study. But participation in exchange for free medical care was not value-free; it was complicated by an overreliance on medicines rather than nutritional or infrastructural changes. In these communities, drugs equaled "good" care; often the use of antibiotics spoke to a family's prestige, their ability to provide. Mothers sometimes felt that because the study's medical aid was "free," it was not good quality, especially when the study doctors refused to give them antibiotics as treatment for diarrhea. Doctors did this both because antibiotics are very overprescribed in Mirpur (as in most places) and because the aim of the research was to look at the gut microbiome, extreme care was taken in giving and keeping track of the antibiotics the children were taking. One doctor told me that misunderstanding antibiotics was a huge problem: "I would prescribe amoxicillin and the mothers would say, 'I know this drug, and it doesn't work!' Then they would go to the unregulated pharmacy where the shopkeepers know the mothers are uneducated, and they will give them some third-tier antibiotic." As a result, mothers might or might not share this information with their FRAs upon the next household visit, fully aware that (reporting) use of these medicines might invalidate their study status. In this study, medical care occupied a precarious place as a form of payment, a kind of gift, and a powerful motivating factor for participation. Much like Biruk describes the "gifting" of soap to Malawian participants in an American longitudinal study on AIDS in Malawi, these sorts of compensations require a "reparative framework for thinking ethics that is responsive not just to project-based parameters but also to the histories and political economy in which projects (and ethics) are situated."[96] My presence and questions further complicated this ethical exchange. Though my translator explained to the mothers that I was an anthropologist and not a physician or scientist, they assumed I was a medical doctor and asked me many, many questions (that I wasn't qualified to answer) about their children's health. And though I had exhaustive IRB approvals for my research, and attempted to explain "informed consent" and their right to refusal, I have no doubt that none of the mothers I spoke to would have declined to answer, out of fear of jeopardizing her place in the study.

Additionally, mothers of case-control children tended to be more dissatisfied and less motivated to continue the study. Because their babies were undernourished when they entered the study, initially the *pushti* packets would make them vomit, and they didn't like eating the therapeutic food every day. After three months of treatment and forcing their babies to eat, many mothers wanted to quit. These mothers were also reluctant to give blood samples, as they believed a blood draw would harm their already weakened children. Blood samples are not technically compulsory but are strongly encouraged by FRAS. One FRA told me she would not enroll a mother who refused to give blood samples: "If they are reluctant to give blood, I will not enroll them, I don't want to enroll them. We explain everything before enrollment, and if they refuse the blood sample I will not enroll them, won't write down their name," even though the informed consent form states "You do not have to give blood if you choose not to." In these ways, the on-the-ground implementation of the Microbiome Discovery Project looks different than is outlined in the Gates Foundation proposals. The FRAS record all their data by hand, in English, in old-fashioned ledgers, using carbon paper to make copies. The supervisor tells me Mal-ED originally requested that the forms be in Bangla, but icddr,b refused, claiming it would be "too complicated." This demonstrates the interesting complexity of "sensitive" development—where the governing body's attempts to be responsive to local needs are eschewed by the local agency in favor of perceived productivity. This information is extensive and includes the sample collection form, a diarrheal episode form, macroscopic/microscopic observation, lab results, and the daily checklist.

In this iteration, microbes are recorded as written information and translated into thousands of forms. The "collaborative" data are dizzying. These forms find their way to the icddr,b Data Management Unit, a gigantic office filled floor to ceiling with jam-packed file cabinets and a staff dedicated to the wrangling of all this information, transcribing it into electronic records. This project is no small endeavor; it is a huge global undertaking that requires a disjointed collaboration between people and microbes, both changing form from biological agents to qualitative data.

The Manifesto

We called it the "manifesto"—we started writing it in January 2011, we submitted it in August 2011, and it was published in *PNAS* in spring 2012, a few days before my and Joe's first child was born.[97] It had thirty versions before I stopped counting, mostly because the increasing number was daunting and discouraging. It was the most difficult writing I had ever done, requiring an extreme expansion of perspective while simultaneously compressing and constricting every academic impulse I had. Dr. Gordon, Joe, and I had conceived of an essay that would outline the terms of our collaboration and serve as a call to arms for others interested in combining the tools and efforts of anthropology and human microbial ecology.

Writing with Dr. Gordon forced me to completely deconstruct my learned disciplinary language, and subsequently it has been very difficult to get back from that place. Just the act of writing together was a monumental struggle—we took for granted at the start that scientists and anthropologists construct everything from sentences to arguments in different ways. I learned the value in having to fight for my ideas, to strengthen them by having to

Notebooks like these were used by FRAs to record all the study data. It continually surprised me that no one ever acknowledged that all the samples, data, and science originated in this very resource-poor place. While the Gordon Lab had cutting-edge equipment, subzero alarmed freezers, and pristine workspaces, the samples started in Dhaka, where electricity was spotty and contamination was unavoidable.

make an argument for them, and not to rely on jargon. There were also some gaps that could not be bridged. A particularly strong argument swirled around the use of the phrase "production of knowledge" by the anthropologist coauthors. In the end, while Dr. Gordon could not agree that knowledge was in fact produced (an understandable perspective for a lifelong scientist), he did concede that entire disciplines have been built around critical theories of knowledge production and that disagreeing with the idea did not discount its value. From our side, we relinquished the inclusion of more nuanced anthropological concepts, such as "ontology," that were too polysemic and bound to bewilder. Some things we fought for, and some things we didn't.

A draft was circulated to lab members, as well as fellow anthropologists, and confusion abounded on both sides. We included a "figure" and a "table," both of which were familiar to me only from grad school as objects of critical analysis, as in Latour's "Drawing Things Together."[98] Our target audience were the readers of PNAS. This publication was Dr. Gordon's choice from the beginning as a journal that would accept a "Perspective" essay (namely, a speculative piece of writing about a proposed future field of study) and one that would appeal to both hard scientists and the more science-leaning social scientists. From his position, he could submit an unsolicited manuscript to many scientific journals of his choosing. This is one of the instances when Dr. Gordon's position was both bane and boon to me as a collaborator. While I would have preferred to submit the article to a more public-oriented or social science journal (and thus not have had to spend so much time on my end defining and explicating anthropology as a discipline), I would never have had the opportunity to publish something in PNAS on my own. Collaboration involves a lot of deciding what level of independent decision-making you are willing to give up, and if what you're gaining is worth it.

In the essay, we made some conceptual and methodological suggestions for studying the anthropology of microbes, and these goals were truly shared. Several years later, I stand behind most of these ideas, especially the methods—even if I have met with more failure than success and am still unsure about how these methods can be operationalized.

CONCEPTS

Investigating concepts of the self and ownership of microbes in specific social contexts

Studying human microbial heritage—of isolated populations and groups—as a complement to cultural, political, and ethnic heritages

Comparing how humans and nonhuman primates and their microbes adapt to different physical and social environments
Understanding how changing lifestyles such as modernization, globalization, food distribution, and migration from rural to urban areas are impacting health and the human microbiome
Analyzing how prenatal and neonatal care is shaped by cultural traditions, and how this affects intergenerational transmission of microbes
Examining the types of relations and networks that are formed within microbiome research: between humans (scientists, anthropologists, and study subjects), between humans and microbes, and between microbes

METHODS

Anthropologists joining scientists who study human microbial ecology and the expressed functions of the human microbiome in the laboratory, sharing and learning methods and concepts—creating a collaborative space, and changing the culture of engagement between the disciplines
Anthropologists working at scientific field sites, conducting ethnographic research at the community level with human microbiome study participants
Scientists and anthropologists designing fieldwork and laboratory experiments, starting the collaborative process upstream, rather than retrospectively
Meaningful consideration of ethnographic data when interpreting scientific results[99]

The collaborator I am now would push harder to problematize the ideas of "isolated populations and groups" (see chapter 5 on race) and work to assert important factors of structural inequality. At the finishing of this manuscript, I have ten more years of experience as a scholar and a fieldworker. I am *much* older and have had countless further interactions with all sorts of different human microbial ecologists. I was able to work with Dr. Gordon precisely because I was a PhD candidate, but at that stage of my educational and professional life, I knew very little about practicing ethnography tactically or how to partner across disciplines. More than anything, writing the manifesto elucidated the extreme difficulty that results from interdisciplinary collaboration between the social and biological sciences, even when the parties bring a high level of willingness, kindness, and flexibility. We not only exposed scientific readers of *PNAS* to the ideas of sociocultural anthropology but, ultimately (and surprisingly), the paper has been used extensively by social scientists as an example of a successful cooperative project. The article has been used in graduate and undergraduate courses that focus on interspecies/posthuman theory, environmental studies, genomics and public health, and medical anthropology. In the subsequent years, I have been contacted by researchers in oncology, demography, geography, biological anthropology, and more who have read the *PNAS* paper and were inspired by the coworking point of view.

When I left the Gordon Lab in October 2011, I was physically leaving the lab, but not the partnership, and would continue to work and write with Dr. Gordon until 2015. In 2011, I made a list of the things collaborating taught me. Social scientists can't think of *science* as a monolith any more than we would think of "culture" in that way. It is unproductive and wrong. People make their worlds, and that means science too. Anthropologists can't be the ethics police; we're not even qualified to do that. Collaborating made me come to question anthropology's motives and outcomes. At times it feels collaboration is nearly impossible. For the most part, scientists don't understand what anthropology is or what it does.

In the lab, I was forced to explain, defend, and practice a more traditional anthropology than I imagined possible for myself. I also learned a lot about DNA sequencing, nutritional genomics, and mice. I am still not qualified to interrogate this work—just like I wouldn't want the scientists to interrogate my research or question my anthropological methods. Gordon Lab scientists do not believe they are creating truth, facts, or anything stable; things for them are in flux, muddled, interconnected. They accept messiness and uncertainty in their work, but definitely not in their publications. Unlike anthropologists, the graduate students and postdocs *practice* their work all the time. Most of the time, science and scientists do have the power, the money, and the upper hand, and that is one reason social scientists need them. Dr. Gordon expected concrete answers but also did concrete things. In his lab, personal entanglements were just as motivating as science.

Much of the time I spent in the Gordon Lab was waiting. Which is presumably what a lot of ethnographic fieldwork is—at least that's how the old story goes. When I discussed things with Dr. Gordon, it wasn't as much putting complex anthropological questions to the scientific work as it was figuring out what he wanted and what I could do as a fledgling anthropologist. In St. Louis and in Dhaka, I wasn't a free agent or an independent researcher. I didn't have the power to wander the lab or the hospital. Whom I talked to, what I said when I talked, what questions I was allowed to raise, and, ultimately, how I wrote about it were all simultaneously made possible and constrained by my relationship with Dr. Gordon. Even as I finish this book, I think hard about every sentence: how anthropologists will read it and what it will mean to the Gordon Lab. By Dr. Gordon's declaration, I was part of the lab family. Family membership granted one a vast amount of access and trust but also opened the door for hazards. With trust comes potential betrayals; with access comes fear of exposure and misrepresentation. What happened after I left St. Louis was a second phase of collaborating that grew increasingly more complicated. The chapters that follow describe what the collaboration looked like in practice, what I worked on in the lab, what I learned, how things are built, and how things fall apart.

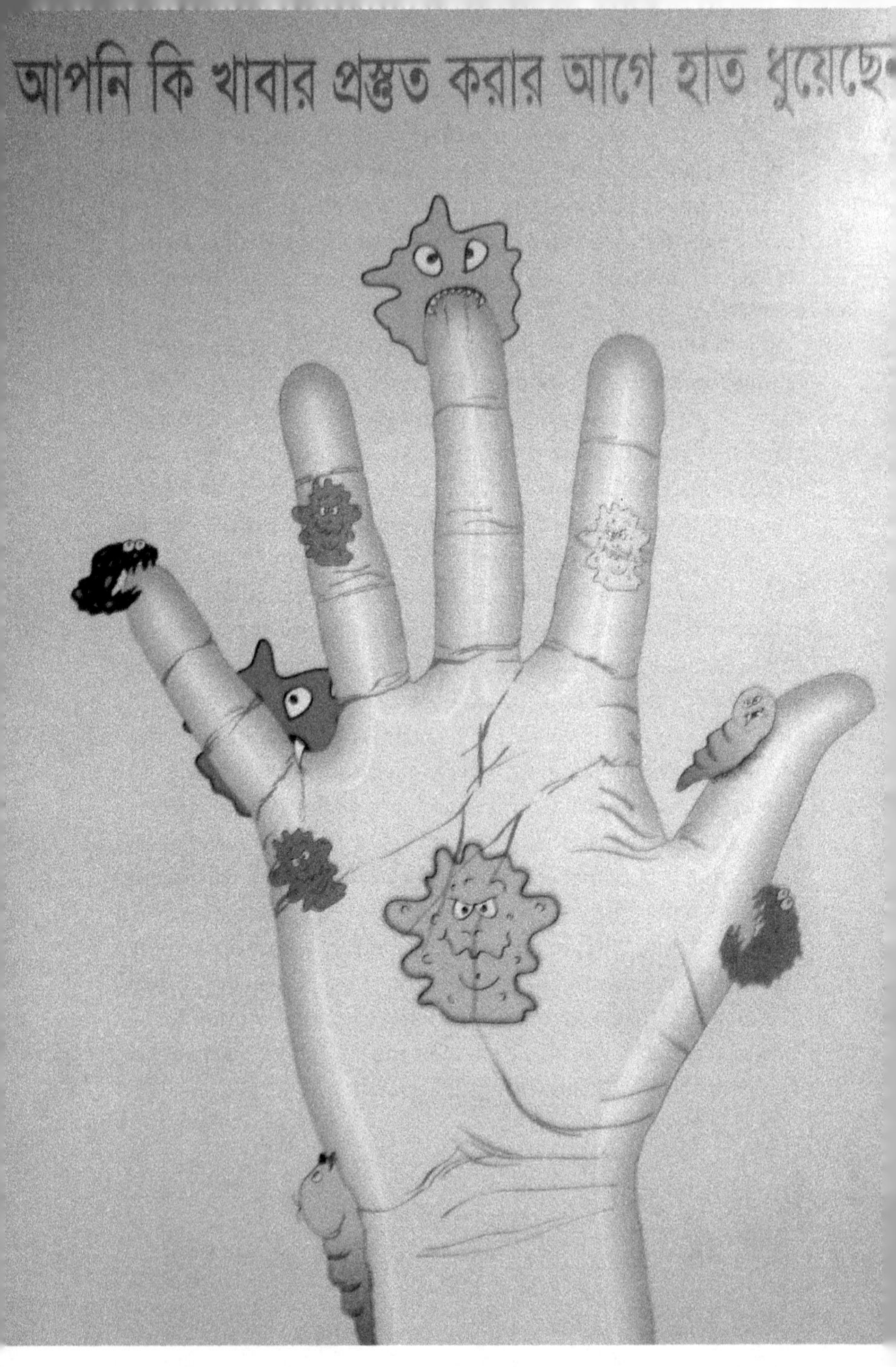

"Did you wash your hands before preparing food?" This sign is commonly seen at both the iccdr,b main hospital and the local Mirpur nutrition centers.

TWO

How to Make a Microbiome

> Microbiologists are fathoming the way modes of thought at the basis of their own discipline have driven biological change. . . . The effects of presuppositions are material, such that the very thing under study has the human history of explanation and intervention within it.
>
> —Hannah Landecker, "Antibiotic Resistance and the Biology of History"

How to Be

I would rather this chapter weren't in this book. I wish I could leave it out. This chapter is the thing I wrote longest ago, probably nine or ten years before *Gut Anthro* was finished. It is a relic of my dissertation and explains the genealogy of science that brings *microbiome* as a concept into being. It references mostly white, able-bodied, hetero-cis-male authors as they tyrannize the fields of microbiology and STS. These are the scholars I was mostly taught, taught to use and cite, my "jurisdiction" as an American graduate student. Revisiting this chapter has made me think hard about what I'm trying to do in it and about my citational practices throughout this book. In *The Body Multiple,* Annemarie Mol asks, how to relate to the literature?[1] As a foundational practice in academia, we must recognize, revere, cite—not only ideas but specific texts and authors. Without a doubt, citational politics are by nature colonizing, white supremacist, and patriarchal and reproduce canons of literature that privilege certain kinds of authors and certain forms of knowledge—as Toni Morrison said, "canon building is empire building."[2] To address this, Mol splits her book in two, discussing the frictions of citations and literature. Unfortunately, we can't all replicate this model, although in their book, geographer Max Liboiron (Michif-settler) does a very good job of using footnotes to work through acknowledgments, relations, citations, and literature as a "part of doing good relations within a text, through a text. Since the main goal of *Pollution Is Colonialism* is to show how methodology is a way of being in the world and that ways of being are tied up in obligation, these footnotes are one way to enact that argument."[3]

Sara Ahmed shows that citational structures form disciplines and that, for instance, all male reading lists, speaker lists, and reference lists (which lead to hiring lists, funding lists, etc.) *screen* out other kinds of bodies and thinkers.[4] And that even fields of feminist theory are dominated by citations of white male philosophers (if I never again read another feminist science studies scholar calling upon Deleuze and Guattari, it will be too soon). In *Living a Feminist Life,* Ahmed makes a move not to cite any "white men" (which she considers an institution, rather than a bunch of individuals) and instead to reference scholars working toward feminism and antiracism: "When citational practices become habits, bricks form walls. I think as feminists we can hope to create a crisis around citation, even just a hesitation, a wondering, that might help us not to follow the well-trodden citational paths."[5] Eve Tuck, K. Wayne Yang, and Rubén Gaztambide-Fernández work with citation practices in *Critical Ethnic Studies* and ask us to "consider what you might want to change about your academic citation practices. Who do you choose to link and re-circulate in your work? Who gets erased? Who should you stop citing?"[6] Following these scholars, throughout the rest of this book, I have intentionally tried to leave out authors and ideas I consider bricks in the dangerous monolith of academic thought. I've looked out for "firsting in research,"[7] where ideas are staked, claimed, and colonized. Like Ahmed, I have tried to foreground feminist, antiracist scholars and deliberately steered away from the conventional canon of STS, actor-network theory, and multispecies theory. I haven't always succeeded. In the end, I've left this chapter in its original form. Is it enough to situate all this hegemonic science and disrupt these scientists as discoverers and innovators, yet still discuss and cite them? Probably not, but it is the best I can do right now.

Microbial Histories

Microbes are, most reductively, organisms that require a microscope to be seen by the human eye. They have been here for 3.5 billion years, dominating the planet and fundamentally changing the chemistry of earth so that humans and every other "higher" life-form can exist. Humans developed the basic skills of life—the translation of genetic code into proteins, lipids, nucleic acid—by inheriting microbial genes. They have been called "our progenitors, our inventors and our keepers."[8] Microbes are present in all three of the domains of life—archaea, bacteria, and eurkarya—wholly making up the first two and representing among the eukaryotes (organisms with nuclei in their cells) with fungi, protists, and algae. In 1990, microbiologist Carl Woese's "Tree of Life" model introduced archaea as an independent domain. Woese's model is generally accepted, though somewhat problematic, because, like previous taxonomic systems, it does not include viruses, which are still scientifically "nonlife" owing to their lack of cellularity and metabolic function, though they do possess other "lively" qualities, such as having genes and reproduction. Microbes' diversity, their populations, and their ability to perform a wide array of metabolic activities to stay alive in any environment far exceed those of all other life-forms.[9] Microbes are everywhere and in everything.

Even though microbes have been on earth 3.49 billion years longer than *Homo sapiens,* one cannot begin to research a "history of microbes" without inevitably ending up with results for a "history of microbiology." The conflation of microorganisms themselves with the scientific study of them is an important point of departure. The invisibility of microbes has required scientific equipment for any human understanding of them; in this way, they have been enacted in tandem with the technologies developed to see them. However, it is crucial to keep in mind that while humans rely on scientific technologies to see microbes, the actions of microbes cannot be reduced to a sociological explanation. Unlike other biotechnological developments, which seek to alter life itself (assisted reproduction, organ transplantation, genetic engineering), most of microbiology has revolved around learning about microbes and what they do. Certainly manipulation and creation of microorganisms are part of the equation, but historically, prevailing microbial science has been about "discovery."[10]

The social science and history of science literature has aptly covered the topics of vision and visuality, exploring the historicity, instruments, imaginations, discourses, and fragility of seeing in science. With the invention of the microscope in the late sixteenth century, the "discovery" of microorganisms is most often credited to Antonie van Leeuwenhoek. From this point, the technologies required to make microbes visible became central to the ways in which humans understand and interact with them. Investigating microbes as causers of illness and food spoilage in the 1700s gave way to experimental tests of the spontaneous generation hypothesis in the nineteenth century; Pasteur's discoveries in vaccinations, fermentation, and pasteurization in the late 1800s; and the development of methods for identifying disease-causing pathogens. Robert Koch, now known as the father of modern bacteriology, created "Koch's postulates" in the late nineteenth century as tools for identifying causative microorganisms and linking them to specific diseases. These postulates depend on two points: the isolation of microbes from their environments and communities and the culturing of them in the laboratory. "Both these emphases have skewed microbiology, and only in very recent decades has alternative work on bacteria as dynamically interacting components of multicellular systems in a diverse range of non-laboratory environments taken hold."[11] Koch's development of microbiological cultures (the method of multiplying microbes by letting them reproduce in predetermined culture media under controlled laboratory conditions) became the universal method for "seeing" microbes. Besides the development of these methods, Koch is most well known for discovering *Mycobacterium tuberculosis,* the bacterium responsible for causing tuberculosis.

Koch's postulates are still considered the gold standard for determining causative organisms in epidemiology, despite subsequent recognition of their limitations. Many scientists I met working in human microbial ecology were frustrated with the continued reliance on these methods, especially in terms of identifying a "species" by how it infected different hosts. "It is ridiculous to say 'this must be organism X because it is acting like Y in humans, or rats or whatever.' Microbiologists cannot let go of their 'species,'" one molecular biologist complained. Following Koch's period, bacteriology came into its own as a scientific discipline, increasingly relying on biochemistry to understand the biological processes of bacteria.[12]

A shift to the tools of biochemistry (such as chromatography, X-ray crystallography, nuclear magnetic resonance spectroscopy, and electron microscopy) pushed the view of microbes from their pathogenic agency (how they grew in a dish and what they infected) to their biochemical makeup and processes. A few microbiological discoveries in the 1940s were responsible for "the dramatic transition from microbial biochemistry to molecular biology and consequently for the movement of microbiology from the periphery to the center of biology."[13] It had been previously shown that the metabolic functions of microbes are very unlike those of plants or animals, that microbes can obtain energy from different sources and with different mechanisms. Yet biochemistry showed a commonality in the unity of the material in all living cells. Studies in the mid-twentieth century demonstrated that bacteria can exchange genetic material (DNA), can mutate spontaneously, and can form enzymes in specific response to the presence of compounds. These studies revealed how microbes use biochemistry in the processes of genetics, unifying the two fields, and were responsible for the "greatest triumphs of molecular genetics in the second half of the twentieth century," having a "profound impact on a range of other disciplines from evolutionary biology to epidemiology."[14] Bacterial analysis was key in the major scientific discoveries of DNA, RNA, and protein synthesis, and research on bacterial genetic systems led to the development of recombinant DNA technology (i.e., molecular cloning), which became fundamental to most modern biological and medical research. The scientific methods for understanding and studying microbes became molecular, shrinking the scale to microbial genes, proteins, and RNA.

> The advent of sequencing technology transformed microbiology's datasets and breadth of knowledge. The cumbersome methods and limited data of early microbial sequencing were rapidly overwhelmed by high-throughput whole-genome sequencing methods.[15]

It is at this point that Woese's research appears. "Although disputed by many taxonomists, especially those outside microbiology, Woese's work made more sense of molecular data and appeared finally to enable a 'natural' phylogenetic classification of bacteria instead of the prevailing phenetic approaches used—however reluctantly—as defaults."[16] Woese started using 16S ribosomal RNA, a type of RNA that is involved in the production of proteins, to trace phylogenetic relationships between different species of bacteria and archaea. All bacteria have 16S rRNA, and it is conserved within a species, so it is used to identify organisms whose 16S rRNA signature is known (or to identify unknown organisms if the signature is not known). The 16S rRNA gene is amplified and sequenced, allowing the taxa of bacteria present in a community to be identified. Yet even as bacterial genomes were being used as a basis for understanding the genetic and biochemical processes of all life, contradictions began to arise that challenged phylogenic histories and taxonomic classification. Genomic data showed phylogenetic inconsistencies between the 16S and other genes used as markers of evolutionary history, "challeng[ing] the practice of equating the evolutionary history of organisms with the history of molecules."[17] Whereas phylogeny (or phylogenesis) is the development and history of evolutionary relationships among groups of organisms (e.g., species, populations), *phylogenetics* is the study of these relationships through molecular sequencing data. Work on three areas of microbial study—biofilms, lateral gene transfer, and quorum sensing—continues to demonstrate that microbes have the ability to defy common biological knowledge about social organization, heritability, and the ways organisms communicate.

Biofilms form when a group of microbes join together by sticking to a surface. While part of the biofilm, the group will secrete a matrix of polymers to hold the structure together, cells will chemically communicate with one another, large suites of each bacterium's genes will be differently regulated, and bacteria within the biofilm will take on specific metabolic functions to benefit the group. In essence, many individual microbes (sometimes across species, or even across domains) operate together as one, plastic organism. Lateral (or horizontal) gene transfer occurs when genes are transferred between organisms in ways other than conventional reproduction. Bacteria can uptake and express foreign genetic material through cell-to-cell contact or through viruses (organisms in a biofilm often practice lateral gene transfer). Bacteria activate this process for survival, to confer antibiotic resistance, and to spread virulence, and it is an important factor in the evolution of microbes, casting doubt on any fixed "tree of life."[18] Quorum sensing is a form of bacterial communication through chemical language—by using different molecules as symbols, microbes can sense the density of their local population ("sense" a quorum) and subsequently coordinate gene expression. Through quorum sensing, bacteria can activate virulence or antibiotic resistance, evade immune systems, create biofilms, and perform a number of other activities based on the circumstances of their immediate environment. Each species has its own language, but all are also multilingual.[19]

Microbiome as Object

The term *microbiome* was coined by molecular biologist Joshua Lederberg to describe our "menagerie of microbes," the ecological community of commensal, symbiotic, and pathogenic microorganisms that literally share our body space and had been all but ignored as determinants of health and disease. Lederberg imagined the human body as comprising three parts: human cells with all their DNA, mitochondria (the energy-producing factories of the cells) with their DNA, and the microbiome—all the bacteria in our bodies and their DNA.[20] There are approximately 100 trillion microbes in and on the human body. To our measly twenty-three thousand human genes, there are possibly billions of microbial genes in the microbiome (which is sometimes also called the second human genome). Microbes (*microbes* mostly means bacteria, but also fungi and archaea) are all over and inside the human body; they populate every area from hair follicles to toenails, mouth, skin, vagina, and intestines.

The human microbiome, then, can be understood as the totality of microbes, their genomes, and the complex interactions (of these microbes, their genetic elements, and their environment) within the habitat of the human body. Commensal gut bacteria, archaea and fungi are suspected to play a role not only in human nutrition but in immune function, metabolism, aging, inflammatory disease, and brain development, among many, many other human health issues. Until now, very little was known about the bacteria in the human body, but in the early 2000s, the scientific purview on microbes completely changed:

> The vast majority of microbial species have never been grown in the laboratory, and options for studying and quantifying the uncultured were severely limited until the development of DNA based culture-independent methods in the 1980s. . . . Culture-independent techniques, which analyze the DNA extracted directly from a sample rather than from individually cultured microbes, allow us to investigate several aspects of microbial communities. These include taxonomic diversity, such as how many of which microbes are present in a community, and functional metagenomics, which attempts to describe which biological tasks the members of a community can or do carry out.[21]

This ecological shift in seeing microbes has brought *microbial ecology* to the fore: the study of microorganisms and of their relationships with one another and with their environments. Human microbiome research has been propelled by new trends in nonreductive biological thinking, such as systems biology, and has become possible wholly through metagenomic sequencing technologies that allow for the study of the invisible communities that coinhabit the human body.[22] Systems biology is an emerging approach in biomedical and biological scientific research that focuses on the study of complex interactions in biological systems using interdisciplinary methods, from a holistic rather than a reductionist perspective. As Astrid Schrader argues in her exploration of the accused fish-killing microorganism *Pfiesteria piscicida*, "how we get to know a species experimentally cannot be separated from the ontological question of what/who they are."[23]

Dr. Gordon is a microbial advocate. More accurately, Dr. Gordon is a human advocate, with a keen, driving interest in the role and power of commensal microbes. On the wall of his office is a framed quote: "Microbes run the world. It's that simple—but they need good spokespeople." He has succeeded, phenomenally, in becoming that spokesperson. Dr. Gordon is thinking about the interconnection of technology with what questions are asked, what microbes are, and what they *do* in human lives:

> Current human microbiome research is addressing questions originally posed by microbiologists more than a century ago, but it is doing so with new and rapidly expanding sets of experimental and computational tools. We are witnessing its inherently interdisciplinary nature, and how it provides an opportunity to forge new alliances, spawn new fields, and transform how we educate ourselves. We are seeing how it can inspire students to expand their understanding of the human condition and to devote themselves to addressing some of the most challenging and pressing global health problems we face this century. It is changing medical microbiology, broadening our understanding of human biology, and providing new insights about the effects of our changing lifestyles. The public is captivated. As such, there is a need for those

> participating in this field to describe their findings clearly, in a sober fashion, to approach bench to bedside translation in a careful, mindful fashion, and to engage in a proactive societal dialogue about the ethical, legal, social, safety and regulatory issues raised by this research. By ignoring the precautionary principle, we risk reactionary responses arising from the absence of a thoughtful, objective framing of issues and proposed solutions, and from inadequate educational outreach.[24]

Of course, innovation and possibility are not the whole of the microbiome story. Like any science touted as the "next big thing," there are doubts and detractors. Microbiologist and social media pundit Johnathan Eisen had a regular feature on his blog *The Tree of Life* called "Overselling the Microbiome," where he coined the phrase *microbiomania* and ruthlessly criticized current publications about microbiome research until 2018.[25] Members of the Gordon Lab, as well as Dr. Gordon himself, were careful about making any ambitious claims about microbiota and urged restraint. When I asked lab scientists what they thought about the current microbiome frenzy, they responded as follows:

> I am quite skeptical of the microbiome being as big as some people are saying. I love the system and the paradigm-shifting conversations that are happening because of the recent resurgence in its study. It's an amazing interaction and definitely plays important roles in human health and disease, however, the effect sizes for many of the implicated diseases and/or health benefits seem to be often over-shadowed by other, more straightforwardly manipulable variables such as diet and/or sanitation levels. There are undoubtedly some great therapies utilizing the microbiome that will arise in the next few years, but it will not be a smoking gun that allows for a cure to all our maladies. (PhD student)

> I think it will always be very difficult to manipulate the human microbiome without food. I believe that the next big thing will be the combination of probiotics and prebiotics to manipulate host health. (PhD student)

> You can tell by the things that are getting published and the journals that they are in that the microbiome is peaking in hype/popularity. I think the microbiome will prove to be much more important to human health that we thought in the past; it has historically been ignored. But the hype is a bit insane. The Yankees couldn't live up to that hype. (postdoctoral fellow)

Following Dr. Gordon's lead, the scientists I knew believed the microbiome was going to be significant to understanding human health, particularly nutrition. But no one was flaunting the microbiome as a sole causal factor of anything; in fact, they didn't even think of the microbiome as a singular thing. In a lab meeting, I once observed several lab members contesting a presenter's study results with the argument that the data were based on what microbiota looked like "for this particular person, on this particular day," and they thought it necessary "to know if the certain bug in one person looks the same as that same bug in another person." In important ways, they saw microbes as conditional and contextual. Dr. Gordon told me that much like microbial ecologies, constantly in flux, "trying to figure out the microbiome and its iterations in reality needs nimbleness and inventiveness rather than standardization." These conditions and contexts, as well as the means through which human microbiota are studied and understood, propel future research directions, raise bioethical issues, and complicate logics of funding.

Community Matters

Since the early 2000s, the development of metagenomics and its associated technologies has begun to address a more ecological approach to understanding microbes within their communities, how they function, and their intricate evolutionary relationships. A *metagenome* is the genetic material from a collection of organisms present in a particular environment, with samples taken directly from, and accounting for, that environment. *Metagenomics* (also referred to as environmental genomics, ecogenomics, or community genomics) pools and studies the genomes of all the organisms in a community and all the functions encoded in the community's DNA. Metagenomic methods are a series of experimental and computational approaches that allow a microbial community's composition to be defined without having to culture its members and are centered around shotgun sequencing, massively parallel sequencing, and bioinformatics. Whereas DNA sequencing is the process of determining the precise order of nucleotides within a DNA molecule, *shotgun sequencing* is a laboratory technique for determining the DNA sequence of an organism's genome. The method involves breaking the genome into a collection of small DNA fragments, which are sequenced individually. A computer program looks for overlaps in the DNA sequences and uses them to place the individual fragments in their correct order to reconstitute the genome.[26] Massively parallel (or next-generation) sequencing is a high-throughput, rapid, low-cost method of sequencing with technologies that perform sequencing processes in parallel, producing thousands or millions of sequences concurrently. Also called computational biology, bioinformatics is the science of using biological data to develop algorithms and relations among various biological systems and to develop and apply methods for storing, retrieving, organizing, visualizing, and analyzing biological data. Because metagenomics looks at whole communities, metagenomic sequencing results in huge data sets, literally hundreds of millions of gene reads. This much biological, genetic, taxonomic, and metabolomic data have never been analyzed, stored, or accessed before, requiring the development of new tools and technologies. In essence, the DNA of an entire community can be sequenced. Metagenomics has transformative implications for microbial science because, like the invention of the microscope four hundred years ago, it provides technological tools for new ways of seeing: "metagenomics has revolutionized microbiology because it offers a window on an enormous and previously unknown world of microorganisms."[27] It also has significance for those who are concerned with the history, philosophy, and application of science:

> Metagenomics requires us to think in different ways about what human beings are and what their relation to the microbial world

> is. Metagenomics could also transform the way in which evolutionary processes are understood, with the most basic relationship between cells from both similar and different organisms being far more cooperative and less antagonistic than is widely assumed. In addition to raising fundamental questions about biological ontology, metagenomics generates possibilities for powerful technologies addressed to issues of climate, health and conservation.[28]

Biogeography is the study of the distribution of species and ecosystems in geographic space and through geological time. Microbiome research has co-opted and redefined this term to describe the vast variety of microbial populations in different bodily terrains and climates, showing that the composition of bacterial communities is determined primarily by body area: "each person can be viewed as an island-like 'patch' of habitat occupied by microbial assemblages formed by the fundamental processes of community ecology: dispersal, local diversification, environmental selection, and ecological drift."[29] Human biogeographical studies scale the entire human body: from the different "climates" of the skin (elbow, colon, nose, vagina) to the Belly Button Biodiversity project. Studies look at how the particular bacteria on tongue and teeth might affect cardiovascular disease;[30] how declining microbial diversity leads to asthma and allergies;[31] how our microbes affect the efficacy of medications; and even how gut microbes influence brain chemistry, moods, and behavior.[32]

In this vast human landscape, the gut contains the body's largest, most varied, and most powerful collection of microbes. By *gut,* I mean the gastrointestinal tract, including all the digestive organs. Besides being the locus of all digestive activity, and the headquarters for the determination of phenotype-related illness like malnutrition and heart disease, the gut is also a crucial player in the immune system. Possessing a huge surface area, the gut is the main place immune cells interact with outside toxins, organisms, and so on. Furthermore, one hundred million neurons (more than the number of neurons in the spinal cord or peripheral nervous system) are embedded in the lining of the gut, making up the enteric nervous system. Sometimes called the "second brain," the enteric nervous system can function independently from the central nervous system, but also stays in close contact with the brain by way of the vagus nerve.[33] Research on the "gut–brain axis" shows correlations between behavior (anxiety, depression, and possibly even autism) and commensal microbes present in the gut.[34] Studies are exploring the gut's role in mood disorders like depression, given that more than 90 percent of the body's serotonin and 50 percent of the body's dopamine are located in the gut.

Gut microbiota research is going in a plethora of directions. Lora Hooper has drawn correlations between diet, immunity, and microbe interactions in gut epithelial cells and mucosa.[35] Martin Blaser has found links between antibiotic use, disappearing microbial species, and conditions like obesity and asthma,[36] while Pete Turnbaugh has shown a direct relationship between our body types and our microbes and how gut microbes can affect the efficacy of some medications we take.[37] Maria Gloria Dominguez-Bello looks at C-sections, vaginal seeding, and the long-term health effects of birth modes.[38] Some are looking at how gut microbiota's effects on immune systems influence cancer or the epidemiological link between cancer and certain bacteria in the microbiome.[39] The company Evelo Biosciences is developing a therapeutics discovery and development platform based on the "human immuno-microbiome" to treat cancer, multiple sclerosis, rheumatoid arthritis, asthma, inflammatory bowel disease, and diabetes through proprietary "monoclonal microbials."[40] The claim is that capsules loaded with a single strain of microorganism would be ingested, and in the intestine, the microbes would commandeer one of the body's routine immunological processes. The HMP, funded as an initiative of the NIH Roadmap for Biomedical Research, is in the second phase of its research, creating integrated longitudinal data sets from both the microbiome and host from three different cohort studies of microbiome-associated conditions using multiple 'omics technologies.[41] Rob Knight is attempting to crowdsource a massive, public microbiota project called American Gut, where participants enroll online and collect and send in their own samples, and the Gordon Lab and collaborators are making big claims about how gut microbes affect malnutrition.[42] The breakneck speed at which this research is advancing was literally too great for me to keep up with here. Every edit of these pages resulted in extensive updating of the scientific literature, and these citations are only representational of an enormous emerging field.

Interspecies, Interdiscipline

Bacteria made the rules for how multicellularity works. Without them, humans would have no immune system; would starve to death, extracting insufficient nutrients from the food they eat; would choke to death on the carbon dioxide they exhale; and would drown in their own urine. Humans cannot, do not, exist without microbes. "Microbes have a proven track record of living in a world devoid of eukaryotes, but multicellular eukaryotes are unlikely to be able to manage in a microbeless ecosphere."[43] Until recently, the biological sciences knew very little about what commensal microbes in the human body were doing. Now microbial thinking is fundamentally altering how biology is scientifically understood, outlined by microbiologist Margaret McFall-Ngai through several key questions:

> How have bacteria facilitated the origin and evolution of animals; how do animals and bacteria affect each other's genomes; how does normal animal development depend on bacterial partners; how is homeostasis maintained between animals and their symbionts; and how can ecological approaches deepen our understanding of the multiple levels of animal–bacterial interaction?[44]

McFall-Ngai calls for a bottom-up reorientation, for the life sciences to recognize that humans, animals, birds, fish, and insects live in a *microbial world.*

A parallel emerging literature on nonhumans in anthropology, sociology, and philosophy of science provides important insights about associations between people, animals, machines, and technology. Companion species, organisms as ecosystems, and the responsibilities that arise from human–nonhuman entanglements are increasingly common themes in the social study of science, some beginning to take particular notice of microbial life. Although the social sciences have a long-standing interest in how microbes affect human social, political, and economic life, the primary focus has been on infectious diseases. Works in the new millennium attempt to address microbes among (and within) us from different scientific and ethical standpoints.[45] This literature urgently promotes a microbial ethos and, to varying degrees, brings scientific explorations of commensal biosocialities into conversation with philosophy, history, and anthropology. In parallel with science, those concerned with the history, practice, and philosophy of science are starting to take on larger questions of nonhuman, biosocial microbial relationships, pushing past solitary concerns for diseases and viruses.

Similarly, on the scene in anthropology, many emerging ethnographies follow Donna Haraway and others in accounting for nonhuman life-forms in social science. Although not a novel concept, this "field" is primarily concerned with bringing long-neglected organisms, such as dogs, primates, meerkats, fungi, and insects, into the equation of environmental studies, STS, and animal studies, with legibly biographical and political lives.[46] Microbes have the potential to explode categories of "species," and even "life," so it seems unproductive to argue for their relevance on continued, humanistic terms. As Haraway points out, "bacteria have never made good species."[47] In *When Species Meet,* she refers to the need to pay attention to all the different players that make up life on earth:

> To knot companion and species together in encounter, in regard and respect, is to enter the world of becoming with, where who and what are is precisely what is at stake. . . . Not much is excluded from the needed play, not technologies, commerce, organisms, landscapes, people, practices. I am not a posthumanist; I am who I become with companion species, who and which make a mess out of categories in the making of kin and kind.[48]

Haraway is still speaking about life, but for her, it is refigured. Life is made by the entangled engagements of parts human and nonhuman, technological and material. This is a continuation of the analytic project Haraway has pursued for decades, a rigorous and ceaseless questioning of who and what make and act in the world. Similarly, biologist Lynn Margulis spent most of a forty-five-year-long career knee-deep in microbes (sometime literally), asserting that microbes are the dominant form of life and that humans have, for millions of years, been constituted by integrated microbial communities: "humans are one kind of self, but we are composed of smaller selves, and we form parts of the more inclusive selves."[49] She is steadfast: we originated as microbes, and they continue to exist in our very cells. For Margulis, everything is unified through a shared genetic history and Gaian superecosystem. She makes no claims that symbiosis is happy, congenial, and cooperative—only that it is necessary.

Margulis, after making revolutionary scientific discoveries of her own,[50] dedicated herself to promoting symbiogenesis as the major factor in evolution, emphasizing the interdependence of all life-forms on earth. *Symbiogenesis* is the emergence of a new organism (phenotype, organ, tissue, or organelle) from the interconnectivity of two separate organisms. Margulis's approach to microbial life is one of unbounded possibility—she sees every cell as a time capsule waiting to be analyzed, holding the stories of human history within its membrane. For her, science explains a genetic, life-dependent interconnectivity between all living things (but the boundaries of "living" are constantly in flux).[51] Margulis occupies a remarkable place in the social study of science, both as an object of investigation and as a participant in critical analysis. She is a complicated figure—some scientific peers have called her theories works of science fiction (perhaps a microbial ethics requires speculative, technoscientific thinking?), while social scientists (perhaps unqualified to assess the microbiology on its own terms) have embraced her. Symbiogenesis is appealing to social scientists as a tool because it focuses on connectivity, engages the nonhuman in bodily ways, and can be used to examine what it means for humans to have coevolved with living and nonliving others. It is used extensively by Myra Hird as a discursive tool to contemplate sex, gender, and selfhood from the perspective of microbes and by Stefan Helmreich as a semiotic framework merging a metaphorical distillation of symbiogenesis with what Heather Paxson has termed "microbiopolitics" to create a theory of *symbiopolitics,* "the governance of relations among entangled living things."[52] Haraway says that Margulis's symbiogenesis "gives me the flesh and figures that companion species need to understand their messmates."[53] All recognize the impossibility of maintaining a division between the biological body and its social life.

Commensalism

Literally meaning "to come to the table together," *commensal* has long been used by anthropologists to describe activities of solidarity, communality, and shared identity, especially to describe rituals and practices of eating.[54] For scientists, commensal relationships (technically, where one interacting species benefits and the other benefits, is unaffected, or is unharmed) are one of many types of interactions between the microbes in the microbiota and their human hosts. Commensal bacteria, or *commensals,* are often described as our friends and partners—the "good guys"—in contrast to microbial pathogens that make us sick, ruin us, or kill us. Human microbial ecologists would agree with Donna Haraway that our commensals are, if not beneficial, at the very least our companions, our messmates, all of us as "a multispecies crowd."[55] I take *commensalism* as a conceptual starting place, as a way to think about the material-semiotic relationship between humans and microbes—a sort of Barad-style intra-action, where "'individuals' do not preexist as such but rather materialize in intra-action. . . . Rather, 'individuals' only exist within phenomena (particular materialized/materializing relations) in their ongoing iteratively intra-active reconfiguring."[56] Through this thinking, "subjects" and "objects" are not sui generis, and so agency is attributed, not to humans or nonhumans, but rather to relational assemblages. This matters to anthropology, and not only in an epistemological way, but in the ways that commensal microbes are acting out, biologically and materially. Can we look to arguments being made, not only for the cultural consideration of the biological, but also for the recognition of biological materiality in cultural anthropology—a symmetry of sorts?[57]

Microbes in/on/making up the human body operationalize social science concepts: Barad's *intra-action,* actor-network theory's *material-semiotic*; assemblages of humans–nonhumans, actors and networks produced through relationships; Mol's *multiple*—a diversity of objects that hang together through various coordinations, more than one, less than many; Haraway's idea of *becoming-with* ("to be one is to become with many"), companion species that make us materially and discursively what we are;[58] and well-developed Indigenous genealogies of nonhuman and environmental relations.[59] The physical (and perhaps social and emotional) functioning of human beings depends on these other actors. Nonhuman and material, natural, yet transmitted and changeable through human action, we are just at the beginning of thinking about and accounting for microbes. Yet these becomings are not value-free, and anthropology is complicit, too, in the making of the microbiome. We've circled back to the start. What are microbes without microbiology, without scientific apparatuses to see what they are doing? How do we know what microbes do? Do we care, only for ourselves?

My translator, Israt, leads me out of the residential Mirpur neighborhoods (or camps) and toward Mirpur Road, the main thoroughfare. Before the Liberation War, Mirpur was its own city and not part of Dhaka.

THREE

Microbiokinships

Right in the Gut

Where is the gut? A corporeal and metaphorical space, the gut is the suggestion of invisible viscera, the most sensitive internal self—soft and vulnerable, near the surface, and barely covered by muscle and skin. The gut is where the outside world comes in and where the inside threatens to be exposed. What we swallow from the outside interacts with our insides in our gut. The gut is the primary place our immune cells meet outside toxins and organisms. Gut idioms show this: to spill your guts is to let the messy, hidden insides come out. Gut expressions range from fortitude and courage (that takes guts; do you have the guts?) to instinct (trust your gut; to have gut feelings) to representing our essential selves (I hate your guts). More than anything, guts are full of shit. Shit tells researchers what microbes are in a subject's gut and what the microbes are doing. The Gordon Lab scientists read the biological microbiome story, but the Bangladeshi samples also give a historicopolitical account, the gendered narrative of a global health project, a kinship story. Human–microbe relationships satisfy definitions of kinship in every way that Euro-American anthropologists (the self-appointed keepers of the kinship concept) have historically qualified it: biological, social, shared risk and responsibility, connections and entanglements, intersubjective and conjoined belonging. More importantly, microbes also take up new (old) kinship formulations by feminist, queer, Indigenous, and Black scholars: oddkin, chemical kin, cohort kin, environmental kin, situated kin, Land/body relations, paracommunicable conditions, kincentric ecologies, Place-Thought, ethical relationality, and relational accountability.[1]

Microbes coevolve with humans, and microbial populations in human bodies are determined by environments and exposures, including family, food and place, health care, race and gender inequities, and toxic pollution. Microbiomes are transgenerational links, disarrangements between different bodies and the outside world. This chapter asserts that microbes are kin—kin that are made of and making environments, across generations. Non-/post-/transhuman theorists have spent decades debating the agency, sociality, and ontology of things like microbes, all the while appropriating and eliding Indigenous cosmologies and scholarship that directly addresses the nonhuman world. Microbial kin evokes Indigenous formulations that necessitate reciprocal, ethical accountability to more-than-human relations. This chapter explores what develops for the biological and social sciences (and for humans) if we call these relations kinships and call microbes our kin.

In microbiome science, human beings are holobionts: single ecological units made up of symbiotic assemblages of human cells and microorganisms. The collection of microbial and human genes in the microbiome is a dynamic and interactive microecosystem that changes over time.[2] Microbiomes, as individual to us as our fingerprints, come into being biosocially—determined by geography, genes, biological states, and social interactions. During all of human history, microbes have been coevolving with us, and microbes pass along familial and social lines. Our relationships with microbes are ancestral, inherited, intergenerational. This chapter attempts to do three things. First and second, it explores how microbes are kin that are made of and making environments, across time. Third, this chapter endeavors to begin to address a pressing need to decolonize ontological studies of nonhumans, particularly microbes. Kim TallBear (Sisseton-Wahpeton Oyate) has been bringing this issue to the STS/anthropology conversation since multispecies ethnography came on the scene in Western academia, contending that "interspecies thinking needs Indigenous standpoints."[3] There is Indigenous thinking, ways of being and knowing that necessarily incorporate "other-than-human" worlds. These ideas and scholars have been summarily erased from STS literature on multi-/inter-/transpeciality. My thinking about microbial kin evokes Zoe Todd's (Métis/otipemisiw) "kin study,"[4] which relies on Indigenous concepts of kincentric ecologies—reciprocal accountability to more-than-human relations. Correspondingly and equally urgent, microbiome science itself needs an antiracist intervention, a task I take up in chapter 5 and elsewhere.[5]

In writing about environmental disasters and structural violence, the Sloughslayers, a group of Black, Indigenous, and people of color (BIPOC) women anthropologists, articulate that "there is no outside," no safety from the propagation of toxins in and out of bodies.[6] Concepts of environment have been exploded by this and other recent work in political ecology, environmental anthropology, geography, and reproductive health. Environments can be the physical world plus toxins; they can be genetic landscapes that span generations and geography. Environments can be global, extraterrestrial, and cellular. Environments can be people. They can be uterine, gut, vaginal, atmospheric. Environments can be exposures—but as Agard-Jones, Roberts, Murphy, and Lee have written,[7] enviro-body boundary crossings are not the same for everyone—exposures are always stratified. Microbiomes are environments of this multiple kind, circulating and confounding the insides of bodies, the outside world, and back again. Engaging energetically with Emily Yates-Doerr's unpacking of the terms *social, determinants,* and *health,* I propose that the parts of the microbiome are not parts at all but come into being instead through material-semiotic indeterminacy.[8] I see microbial kinships as bio-socio-enviro-exposo amalgamations, not separate from the science that defines microbiomes, voids structural contexts, helps and harms. The Sloughslayers cite the Lakota philosophy (present in ways across many Indigenous, First Nations, Inuit, and Métis worldviews) of *all our relations,* accounting for the embedded relationality of environment, people, nonhumans, ancestors, and descendants.[9] Underrepresented across disciplines, when this thinking *is* used in the humanities and social sciences, it is often as philosophical novelty expunged of commensurate contexts of land sovereignty, genocide, or politics. When it is used in the sciences, it is without Indigenous input, practices, or acknowledgment of traditional paradigms.[10] Not surprisingly, important innovative and engaged research comes from queer studies, critical race/ethnic studies, Indigenous studies, environmental studies, and geography, instead of anthropology. I don't want to be part of a legacy of appropriation and intellectual theft, nor do I want to be part of an erasure. As I begin to learn from this Indigenous scholarship, I am far from qualified to use it; but I cannot leave it out. Its representation here is perhaps spotty and incomplete, requiring more study, more reading, a seeking out of Indigenous collaborators and teachers, and learning when knowledge is not mine to take.[11] Because neither my academic PhD nor my professional training in anthropology and STS included any of this literature, or accounting for these ways of knowing, I recognize that there is a lot of work to do. I wish this engagement was not contained only to this chapter, as it should reach throughout the book. As a white scholar, I am trying to interrogate my own privilege, colonizing practices, and failures to start to mitigate legacies of harm. I would like to recognize the work done by Indigenous thinkers, to consider how microbes figure and can be used, not to disregard the anthropological commitment to human cultural worlds, but to redefine those worlds by reckoning with nonhumans—microbes as not only good to think with but good to *be* with.

Study Sights

During my first week in the Gordon Lab, a frozen Styrofoam cooler was delivered by DHL; I helped a grad student fish out hundreds of two-inch, hand-labeled plastic vials. Filled to the top with a mostly orange, squishy substance (stool samples from Bangladeshi babies), the vials were immediately sorted by discrete study subject numbers and placed in one of the lab's specific-for-human-samples, always-locked −80°C freezers. The scientists wanted to learn how gut microbes are affected by diet, genes, and environment and whether microbial communities can be intentionally and durably altered to treat malnutrition. Shit was a means to an end, for me too. Ethnographically, I tracked where and who the shit came from.

Though the money came from the Gates Foundation Mal-ED consortium, icddr,b provided these samples, as well as the staff, scientists, FRAs, facilities, and equipment for the collection and initial analysis of the microbes in the Microbiome Discovery Project—and all the research labor. The organization enrolled the families; developed relationships; established trust, took blood, urine, milk, and shit; and collected data. As discussed in chapter 1, icddr,b has a powerful, almost paternalistic presence in Dhaka generally and Mirpur specifically. A few months after I helped unpack those samples in the lab, I blearily arrived at Dhaka's Hazrat Shahjalal Airport in the muggy dark of 4:00 A.M. It was the middle of the Eid al-Adha festival, and I didn't know yet that the airport and streets were empty because most of Dhaka's population was visiting family in the rural areas for the holiday. As the sun rose over the bamboo scaffolding of half-finished high rises, the people were quiet, and birds and insects filled the sky. As I was the first Gordon Lab member to travel to Bangladesh, I was meant to be the lab's anthropological envoy and its eyes and ears on the ground. The collection of ethnographic data would help identify and characterize family structures, food habits/distribution, childcare, and traditional health practices. It would also deliver "culturally relevant" data on eating practices to help scientists create Bangladeshi diets for gnotobiotic mice. Over the next six months, I was to generate an in-depth understanding of how household interactions (touching, feeding, sharing food and utensils) influenced the constitution of microbial ecologies while photographically documenting conditions of sanitation, food, homes, and evidence of Westernization. In the Gordon Lab, "Westernization" meant new forms of food acquisition, availability, and preferences that accompanied economic development. Lab scientists described a Westernized diet as access to nonindigenous, processed foods or foods with a novel nutritional makeup. Westernization was understood to have pernicious effects on health.[12]

Enteric disease is a daily and endemic problem in urban Bangladesh—the convergence of extreme population density; a decrepit sewage system; no government-supported sanitation infrastructure; and high concentrations of pathogens in drinking water, vegetables, and street food. As Michelle Murphy (Winnipeg Métis) has said, "we are all here already more-than-bodies. This material, not metaphorical, entanglement in environmental violence is a condition of being alive today."[13] Environments and microbiomes in Dhaka are made through coinciding political, humanitarian, colonial, technological, climatological, and capitalist regimes.

The icddr,b field site in Mirpur (a subdistrict of urban Dhaka) is five square kilometers in area and houses fifty thousand residents, most of whom are rickshaw pullers, garment workers, and mothers. To these Bangladeshis, gut microbes are not kin. Microbes cause chronic diarrhea, small babies,[14] and ubiquitous NGO public health intervention. Or perhaps microbes here are the type of kin that harms and kills, inescapable microbial entanglements that are not utopic cross-species companionships. As Dow and Lamoreaux write, "kin relations are not always strategic engagements or willful encounters. Kinship might also be unintentional, unwanted, or unknown entanglements with other beings. . . . Situated kinmaking is not only about strategic alliances but also about the destructive elements of relations."[15] Social disruptions can cause microbial perturbations, and microbial perturbations cause social disruptions. Because microbiomes create gut environments and are also created by socioeconomic circumstances, they are kin situated across scales.

In Bangladesh, generations of families have been health and social science research subjects of icddr,b for more than fifty years, and study participation has become a circumstance of everyday life. The Microbiome Discovery Project was circumscribed by what V. K. Nguyen has called "government-by-exception," the nonlegal rule of NGOs, hospitals, and research institutions that directly governs the lives of populations through "emergency" health interventions.[16] A significant consequence of Bangladesh's political past—centuries of British colonial rule, followed by Pakistani military occupation, and national freedom only in 1971—is substantial economic dependence on interposed international organizations. Relentless famines, droughts, cyclones, and floods in the interceding years have created extensive further involvement by relief and development NGOs. In her book *Threatening Dystopias,* geographer Kasia Paprocki posits that this history has resulted in a fusion of development and climate change—development politics and rhetoric in Bangladesh are now viewed as part of climate change adaptations. Changing climate has led to rural agrarian failures and large-scale employment migration to cities, resulting in increasing export commodities (like shrimp and textiles).[17] New narratives frame climate change, then, as presenting opportunities for development and economic growth, with NGOs intimately involved. In 2008, a staggering twenty-six thousand NGOs were formally registered with the Bangladesh NGO Affairs Bureau. As icddr,b compensated for the state shortfall in social services, health care, and infrastructure, it also became a globally prominent institution for accumulating a massive amount of data that measured, quantified, and experimented with populations over generations.[18] icddr,b's engagement in the Mirpur community is persistent and complicated. The institutional strategic plan includes goals like "develop and promote use of innovations for Bangladesh and the Global South" and "invest in our people."[19] These goals were large-scale but played out continuously in local contexts. In my fieldwork at icddr,b, simultaneous to this mission I saw Bangladeshi researchers forced to capitulate to American institutions; class, ethnic, and gender disparities; and troublesome issues of research consent.

As I learned about practices like cohabitation, nursing, cooking, and feeding, while I also learned about the science, I saw how gut microbiomes were amalgamations of everyday life that shape the evolutionary cohistories of humans and microbes. I continuously tried to foreground the structural causes of malnutrition—poverty, household job insecurity, inadequate health care, failures of infrastructure, toxic exposures—while the lab studied diarrheal infections, gut microbiota, and feeding. My partnership with the Gordon Lab required recording and uncoupling social, political, and environmental conditions from the bodies embedded in them. This chafed against my ethnographic sensibility, as did endless requests for quantified data that I couldn't and wouldn't provide.

The plastic tubes that mothers filled with their babies' shit each morning traveled and transmitted Mirpur both biologically and conceptually into the Gordon Lab. All the samples would be distributed into smaller and smaller tubes, eventually prepared for RNA extraction (to see what microbes are there), or into solutions for colonizing and humanizing germ-free mouse guts (to see what they do). For the Gordon Lab, the microbiomes contained in these samples said a lot about the environment of the donor, their kin, and their world. These researchers were interested in how microbes moved between and in people and spaces. Like Elizabeth Roberts's environmental science collaborators, "they are interested in what passes from mothers to children, with what effects, and from environments into bodies, with what effects. They are interested in transmission."[20] Roberts describes (reproductive) anthropologists who deal in life circumstances that shape biological health as *environmental transmissionists.* This is also an apt way to describe scientists and anthropologists working on the microbiome.

Exposome is an environmental science term that refers to all environmental exposures that are not genomic. In response to the exposome, social scientists have developed the framework of the "socio-exposome," warning against the molecularization of inequalities and pushing for the inclusion of sociopolitical conditions to be considered in environmental exposures:

> Environmental scientists coined the term "exposome" with the goal of inventorying and quantifying environmental exposures as precisely as scientists measure genes and gene expression. . . . Our socio-exposome framework blends insights from sociological and public health research with insights from environmental justice

> scholarship and activism. We argue that environmental health science requires more comprehensive data on more and different kinds of environmental exposures, but also must consider the socio-political conditions and inequalities that allow hazards to continue unchecked.[21]

Instead of looking at how environments are turned into the biology of the body, human microbial ecology looks more widely and across individual development at how microbial populations are constituted by worldly environments—early environmental exposures define gut microbial communities. Human–microbial bodies are socio-exposo bodies.

The Environment Multiple

Nutrition Centre assistant director Shamshir Ahmed told me he was proud of icddr,b's relationship with the community, that people trust them. This Mirpur 11 site has served as an icddr,b hub for community-based biomedical and socioeconomic interventions since 2000, and it is where the microbiome study population resides. By the United Nations definition, Mirpur 11 is an urban slum, denoting an area characterized by substandard housing; overcrowding; and high rates of poverty, illiteracy, and unemployment, along with high rates of disease due to unsanitary conditions, malnutrition, lack of basic health care, and inadequate access to safe water.[22] The residents of Mirpur spoke of sections, camps, and communities—of course, *slum* is a phrase used only by doctors, program officers, and foreign donors and never concedes to the colonial, capitalist violence that creates the conditions of place. The toxic exposures in Mirpur were many: open drains outside homes and pit latrines inside them, maintained and cleaned by residents because there was no functional city infrastructure for sewage or garbage collection; smoke from wood cooking stoves for homes without electricity; copious pesticides and pathogens in food; air pollution from industrial manufacturing and an enormous number of motor vehicles. As developed in the critical framework by Max Liboiron (Michif-settler),[23] Reena Shadaan, and Michelle Murphy, "pollution is colonialism."[24] Doing fieldwork in Belize, Amy Moran-Thomas came to describe chronic health issues like diabetes that were entangled with exposures "as para-communicable—chronic conditions that may be materially transmitted as bodies and ecologies intimately shape each other over time, with unequal and compounding effects for historically situated groups of people."[25] In the same way, the placeness of Mirpur, enteric disease, and malnutrition through microbiomes are para-communicable. What microbes are and do in this place are influenced by ecologies of poverty and the material conditions particular to this place.

The residents of the five camps in Mirpur are designated by icddr,b as "poor" or "ultra-poor." There was no consensus about these terms, even among the staff. Fieldworkers were navigating definitions of (environmental, infrastructural, employment) poverty prescribed by global health organizations and were sensitive to what those designations might mean to a (white, American) representative from the Gordon Lab. They were familiar with the performativity required of them by foreign donors and researchers but also tried to hold stable their commitments to the Mirpur communities. *Poor* and *ultra-poor* came to simultaneously describe environments and people, and eventually their corresponding microbiomes. Most of the "poor" homes in this community are one-room, ten-by-twelve-foot structures made of concrete floors and tin roofs with bamboo or tin walls. The average number of household members ranges from four to ten people, depending on the number of extended family members living in a single dwelling. The average income for families in this community is between $50 and $130 per month. The community in Mirpur 11 is primarily Muslim and speaks the native language of Bangla (Bengali). There is one large market, several smaller ones, and three Mal-ED Nutrition Centres. According to the quarterly census taken by FRAS at the field office, the study area consists of 9,250 households.

Drinking water in Dhaka is notoriously and extraordinarily contaminated.[26] Most Mirpur households have government-supplied water, which is piped into reserve tanks for use by several households. Water is retrieved from the reserve tank by a tube well pump. This water is not safe to drink without boiling because it is contaminated with sewage, agrochemicals, oil spillage, solid waste, and untreated industrial effluent. Even then, boiling does not rid it of high levels of arsenic and other heavy metals. While families are encouraged (by NGOs, local government, and icddr,b) to boil water before consuming it, Mirpur families must contend with unreliable gas service and the high cost of fuel, then prioritize best uses for their limited access to stoves. Many mothers said they chose to use their one or two hours on the stove to cook their families' food for the day instead of the time-consuming task of boiling huge pots of water. Consequently, the lack of clean water for drinking, cooking, and washing has profound effects on health and on gut microbial populations.[27] FRA Roji told me, "Water is one of the most important things to teach about—impure water contains germs, and this is one of the main factors in diarrhea. They [mothers] know to boil their water, but it is difficult for them because they do not have sufficient systems in place. This is the most important thing to prevent disease." This shows how external and infrastructural exposures figure into how microbial populations are constituted in familial guts. The disconnect between health education, NGO intervention, and community "compliance" is fraught with practical obstacles. The social determinants of health framework here is concerned with illness caused by contaminated water but stops short at boiling directives. Studying the environmental and kinship complexity of the microbiome also "asks how health comes to matter to people in their lives, undertakes interventions in conversation with the people whose lives are impacted by public health's interventions."[28] Mandates to boil water aren't fulfilled when boiling water involves a series of intricate decisions about familial happiness and health, risking illness to eat.

In microbiome studies, what environment *is* is complicated, flexible, and multiple: environments can be exposure to toxins, behaviors, or organisms; each version of it can be seen as enacted through the scientific and political processes that make it a knowable object.[29] As Hannah Landecker shows, what emerges when we start to monitor ecologies made by science is that "the bodily condition bleeds into the environmental condition."[30]

When boiling water became untenable, Mirpur families often let water sit out for a day to "sanitize" it, they told me.

There is also a correlative issue sometimes raised in human microbial ecology—is the microbiome part of the organism or part of the environment? Anthropologist Harris Solomon asks a similar question, "what is body and what is environment?" while developing his concept of *metabolic living* through his fieldwork in the Mumbai neighborhood of Bandra.[31] Solomon is working out ideas of temporality and location in medicalized metabolic disorder (namely, obesity), thinking about technoscience, feeding, political interfaces, and where bodies begin and end:

> Abiding ways to ask about the inside/outside boundaries of the body and the world often rely on the supposition that bodies in the same space are sharing the same world, made up of the same objects of concern. Yet the sciences and lived experience of metabolism unseat that supposition. They demand a detailed look at the situated, absorptive interface of bodies and surroundings in which a condition like metabolic illness lodges inside a body.[32]

Metabolic living maps onto an anthropology of microbes in numerous ways; the dominant role of microbes in metabolism also disarrays tidy understandings of in and out. The microbiome has been elevated to the status of "metabolic organ," crucial for glucose and lipid metabolism, glycemic control, fat storage, gut dysbiosis, and endocrine function.[33] The scientific imaginary revels in the idea of a human-bacterial superorganism with a colossal metabolic capacity, especially the possibility of finessing the function of the microbiome to cause altered metabolic states for humans. Accordingly, the Gordon Lab deals exclusively in metabolic disorders, ranging from malnutrition to obesity, and all their work pivots on the microbiome's manipulability and metabolic power. So the body multiplies, becomes a socio-exposo-microbiome body, simultaneously multiple and partial, a matter of scale.

One of the more compelling anthropological concepts of the last thirty years, Marilyn Strathern described her concept of "changing scale" simply as "switching from one perspective on a phenomenon to another"[34] but used scale to address the anthropological challenge of dealing with the incommensurable amplitudes of our data, analysis, and comparison. At the same time, scales are fabricated and parameters chosen from a plurality of wholes. Each part itself is not a fragment but a whole made through connections. The size of something never reduces its complexity. "Partial connections" does not refer to a partiality but to "relations through partition."[35] As Annemarie Mol has now famously summarized Strathern, "more than one, less than many . . . the multiplication [of a thing] and the coordination of this multitude into a singularity."[36] Regardless of the multiplying of worlds, the intellectual coordinates of analysis remain the same. Everything is fragmented; everything is connected; any part of one thing can be part of something else. And this is precisely what a microbiome can be—trillions of individual organisms inside, but also constituting the human body, not separable parts, but not an easy whole. And not only our biological selves but our historical, ancestral ones—the bacterial parts of our genomes telling a story of migration, kinship, the food we ate, and who we slept beside. Human–microbe commensality as a primary partial connection. However, all these ideas had been well established in Indigenous cosmologies and concepts long before Strathern, Mol, and me.

Metagenomics, the previously mentioned primary tool for studying the microbiome (also referred to as environmental genomics or community genomics), is the main tool for studying the microbiome. A metagenome is the genetic material from a collection of organisms present in a particular environment. In microbiome science, environments can be "different socioeconomic, geographic, and cultural settings"[37] that transmit or change human microbiomes. For environmental anthropology, environmental/exposure boundaries are sometimes necessary for survival, and exposure burdens are not equitably shared.[38] As Landecker has surmised, food can be an environmental exposure, especially in relation to the metabolic activity and gene expression of microbes.[39] Importantly, environments can be the hands, mouths, and intestines of humans. For scholars like Michi Saagiig Nishnaabeg author Leanne Betasamosake Simpson, drawing distinctions between people, nonhumans, and environments is part of a politics of extraction: "You use gender violence to remove Indigenous peoples and their descendants from the land, you remove agency from the plant and animal worlds and you reposition aki (the land) as 'natural resources' for the use and betterment of white people."[40] What are the implications for racist capitalism, land sovereignty, and heteropatriarchial power if humans are not bounded selves in bounded environments? In translatable ways, these same extractions show up in Dhaka: the double colonization of British and Pakistani rule, the continuous capitalist rule/gendered violence of the garment industry, and the complicated reign of development NGOs.

We are not freestanding units, not only because we are made up of microbial kin who are made of environments but because *we are also making up the environment in which microbial genes are activated*—microbial genes that constitute 99 percent of all the genes in our bodies. In her examination of what antibiotic resistance has done throughout time and throughout bodies, Landecker fittingly comments, "Individual therapies targeted at single pathogens in individual bodies are environmental events affecting bacterial evolution far beyond bodies."[41] While we have between twenty thousand and twenty-five thousand human genes, there are more than forty-six million microbial genes found just in the oral and gut human metagenomes.[42] All this genetic diversity could be bacteria's ability to evolve their DNA in response to changes in the host environment. Researchers speculate that what food people eat, what medicine they use, and what their environmental exposures are can cause microbes to rapidly change their DNA. The environment of our bodies becomes a site for our microbial partners to experimentally evolve their/our genomes.

Gordon Lab studies have found that certain human interactions have a distinct effect on microbial populations in babies, and those who cohabitate are more likely to have similar microbes than those who are genetically related. "Differences in social structures may influence the extent of vertical transmission of the microbiota and the flow of microbes and microbial genes among members of a household. These latter observations emphasize the importance of a history of numerous common environmental exposures in shaping gut microbial ecology."[43] This study considered mothers' intimate practices—birth, touching, feeding, care: measurable environmental exposures. This is an important scaling that factors in human-microbe kin, emphasizing relations in the complexity of living systems and interweaving multiple forms of environment.[44]

Unmaking Kin

In their edited volume *Making Kin Not Population,*[45] Clarke and Haraway link environmental and reproductive justice and call for the creative kinmaking of "oddkin" rather than "biogenetic" kin. The book's thrust is toward what is called "multispecies reproductive justice," which connects the survival of all of us on earth with a necessary rethinking of kinship. Other contributors to the volume, Ruha Benjamin, Yu-ling Huang and Chia-Ling Wu, and Kim TallBear, write about the shapes and possibilities of kin: from relations with those killed by racist police action, to intergenerational relationships, to families that survive despite colonial settler violence. But as Subramaniam[46] and Dow and Lamoreaux[47] deftly show in their analyses of *Making Kin,* Clarke and Haraway problematically center "nonnatalist" population politics, strangely reifying the social–biological divide and depoliticizing legacies of overpopulation narratives. In direct opposition, Michelle Murphy's chapter argues against population as a biological category and outlines the harms the concept does to women of color, Indigenous women, and poor women worldwide.

These are crucial points for me too; as the science of the Gordon Lab enables the making of kin with microbes, care must be taken not to erase the people from which the microbes came. In the excitement to trace relations between humans and microbes, we shouldn't render invisible the relations between mostly white American scientists (and anthropologists) and Bangladeshi mothers and babies. Perhaps in the rush to make kin, we should instead consider all the ways in which we are unmade.

In "Black AfterLives Matter: Cultivating Kinfulness as Reproductive Justice," her chapter in *Making Kin,* Benjamin discusses kinships beyond the biological with "materially dead/spiritually alive ancestors in our midst. . . . Materializing meta-kinship that exceeds biological relatedness continues to take many forms."[48] She calls STS and nonhuman studies to task for too few works on multispecies justice, for too little attention paid to "immaterial actants inhabiting the ancestral landscapes"[49] that matter to Black and Indigenous people. Moreover, TallBear shows how adaptations beyond traditional conceptions of family persist in the face of violent settler-colonial attempts to unmake Indigenous kinships in her chapter "Making Love and Relations beyond Settler Sex and Family." She makes the powerful point that kinships are themselves relational—some survive because others are destroyed, and white kin flourish only through the decimation of kin of color (which includes those other-than-human):

> in short, white bodies and white families in spaces of safety have been propagated in intimate co-constitution with the culling of: black, red, and brown bodies and the wastelanding of their spaces. Who gets to have babies and who does not? Whose babies get to live? Whose do not? Whose relatives, including those other-than-humans will thrive and whose will be laid to waste?[50]

As environments are multiple, kinships are multiple too. In "Toxicology and the Chemistry of Cohort Kinship,"[51] Janelle Lamoreaux gives a short but thorough history of anthropology's legacy of kinship, pointing to new directions where nonhumans become kin. I am interested in expanding this engagement to thinkers who reaffirm the Indigenous cosmologies already doing this work. The ontological turn led many Euro-American scholars to write about how the biological and social are unavoidably coconstitutive and people, nonhumans, and environment come into being through their interactions. Rarely has ontological theory, new materialism, or non-/post-/transhumanism cited or incorporated Indigenous scholarship on these topics. When talking about microbes, no one from Latour to Haraway to Helmreich has worked with Indigenous ideas of relationality and nonhumans. This suggests Lauren Beck's concept of *firsting*—how historical narratives are discourses that always privilege the colonizer.[52] Liboiron (Michif-settler) takes up firsting in research, asking, why is it an indicator of good research to be "the first" to write about an idea, to be a discoverer? Liboiron says firsting is "imperialist and colonial in nature, using language of priority, exploration, discovery, and uniqueness in a way that erases other people and forms of knowledge."[53] As Liboiron points out, of course, nothing is "discovered" that wasn't already known to the Indigenous, existing, or local populations. The problem with firsting in research is that it perpetuates colonial power, erasing forms of knowledge and experience that were already there and laying authoritative claim to ideas. White scholars (including me) have nothing "new" to say about microbes and humans' relationships with microbes that hasn't already been explored by Indigenous thinkers.[54]

Scholars Dwayne Donald (Papaschase Cree), Nicholas J. Reo (Chippewa), and Enrique Salmón (Rarámuri) all develop concepts keenly important to thinking about microbial kin. Donald's *ethical relationality* is based on the Cree teaching of *wahkohtowin*:

> how our histories and experiences position us in relation to one another, and how our futures as people in the world are similarly tied

> together. It is an ethical imperative to remember that we as human beings live in the world together and also alongside our more-than-human relatives; we are called to constantly think and act with reference to those relationships.[55]

Working with the Rarámuri concept of *iwigara,* Salmón writes about *kincentric ecology,* "an awareness that life in any environment is viable only when humans view the life surrounding them as kin. The kin, or relatives, include all the natural elements of an ecosystem."[56] Reo's concept of *relational accountability* is a directive about how to do research: "researchers are responsible for nurturing honorable relationships with community collaborators and are accountable to the entirety of the community in which they work, potentially including collaborators' more-than-human network of relations."[57] Microbiomes are relational processes, enacted and inherited through modes of birth, food prepared and fed to families, participation in legacies of scientific research. Microbial kinships amalgamate genomic inheritance and environmental interaction across generations. Microbial kin is a place to start engaging germane Indigenous formulations of relationality and environment that necessitate reciprocal, ethical accountability to more-than-human relations.

Microbes have been on earth for 3.5 billion years, and humans developed the basic skills of life—the translation of genetic code into proteins, lipids, nucleic acid—by inheriting microbial genes: "microbial inheritance, where a proportion of our microbial inhabitants and their genes are passed between family members and once acquired are retained throughout life, parallels to a degree the inheritance of our human genes."[58] Increasingly, gut microbiome studies in primates and humans show that host genetics plays a consistent role in the heritability of microbes, but that inheritance is entirely environmentally contingent.[59] Microbiome scientists are dealing in kinships between people and microbes that are dependent on material and social environments.

Anthropologists have long studied kinship systems based on shared bodily substance, direct descent, concrete networks of relationships, and forms of reproduction. Studies of kinship have descended from bloodlines of biopolitics and biosociality to focus on technologically transformed forms of reproducing human bodies. Sarah Franklin talks about the future of kinship as a future of biology as technology and has been instrumental in transforming the ways in which anthropologists talk about kin.[60] Surely, assistive reproductive technologies, organ donations, prostheses, robotics, and transgenics hybridize bodies with technology, creating kinships through technological action. Microbial kin occupy a slightly offset fluctuating temporal and technological space. Made through metagenomics, these kinships show microbes and humans working together, for the time being, on successful coevolution. Is it so hard to imagine microbial roots anchoring our family trees?

Take this kinship story. A microscopic photograph of *Bifidobacterium longum* (subspecies *infantis*) shows furry-edged, coral-like arms growing in every direction. *B. infantis* is an anaerobic, rod-shaped bacterium that is present in the intestines of most humans, animals, and insects. We get these microbes from the person who gives birth to us, although no one is sure how, since adults seem to harbor the closely related *B. longum* in our guts, but not the specialized strain of *B. infantis.* The soft branches of *B. infantis* reach out and grow in our guts soon after birth. To be more specific, babies born vaginally seem to immediately acquire *B. infantis* and other microbes. Babies born by cesarean section are shown to have less diversity in this initial composition of their gut (and skin) bacterial communities and in particular lack *Bifidobacterium,* which has shown to result in childhood immune, allergy response, and intestinal disorders.[61] Before long, *B. infantis* is the dominant species in the gut ecosystem. This is good for biological survival, because these bacteria are crucial to the proper functioning of a human gastrointestinal tract: maintaining gut homeostasis, allowing for healthy digestion, inhibiting the growth of viruses and harmful bacteria, and stimulating the immune system of its host. *B. infantis* does all this fueled by complex sugars and has developed a group of genes that enable it to digest the human milk oligosaccharides, or HMOs, that make up 20 percent of breast milk. Because human babies cannot digest HMOs, evolutionary biologists and microbiologists have long suspected that these complex sugars are in breast milk to nourish *B. infantis.* Using new methods in genome sequencing, metabolomics, and chromatography, scientists are now substantiating these suspicions that milk serves as prebiotic for this specific, beneficial bacteria. Further, they've found that mothers with inactive alleles of the FUT2 gene (termed the "secretor" gene) produce an abundance of fucosylated glycans in their breast milk for *B. infantis* to consume. The babies of these mothers have thriving communities of *B. infantis.* Recent research has shown that even the introduction of solid foods and new diets cannot disrupt this relationship; as long as the baby breastfeeds, the gut community remains relatively undiverse, with *B. infantis* at the top of the heap. But this bacteria's heyday is short-lived—after the child is weaned, the microbe no longer survives in the gut and is replaced by *B. longum,* which does not have the ability to metabolize breast milk HMOs. During the 160 million years mammalian nursing has been evolving, postpartum bodies have been creating food for their children as well as food specifically tailored for the microbes in their babies' guts. Breastfeeding humans feed one plus one hundred billion kin.

The scientific understanding of the interaction between *B. infantis* and its human hosts is rife with biological intention; human breast milk biochemically evolves nutrients and bioactive components that best support microbial partners, and breastfeeding parents develop genes to support the production of these glycans. Simultaneously, *B. infantis* develops the seven hundred unique genes required to make it the only microbe able to metabolize the HMOs the baby can't digest.[62] Mothers want to maximize their babies' chances of short- and long-term survival; babies want to live and thrive; microbes want to dominate an ecosystem without competition for resources. Finally, through bodily actions, the newborn physically acquires vaginal and fecal microbes during the birth process, followed by further microbial population through skin and oral contact through nursing, kissing, and touch. Each actor in the system has evolutionarily adjusted to fit into the complex interplay. As long as humans have existed as mammals that produced milk to feed infants, microbes have been populating those infant guts and genetically adapting to the nutritional composition of breast milk. Environment also matters in this relationship: what the physical landscape looks and feels like; what type of weather rains, floods, or beats sun down upon them; what pollutants, toxins, plants, or animals exist; what foods were available and eaten.

Historically, anthropologists have investigated social and biological associations that constitute relatedness and the cultural specificity and fluidity of form and meaning of these relations. What is hundreds of thousands of years of committed cooperative evolution and the simultaneous development of intimate social practices if not transgenerational kinship? To be clear, while many social scientists and philosophers are excited about these relationships because of the connection and companionableness they insinuate, microbial kin are not just happy-go-lucky messmates.[63] Once a microbial ecologist told me, "Commensal microbes are friends until they aren't. There's no such thing as a 'good' or 'bad' microbe." The merit or menace of microbes is entirely dependent on where, when, and how they are situated.

Mother of Microbes

In Mirpur, nineteen FRAs worked full-time with the mothers enrolled in the microbiome project (impressively, these nineteen FRAs visited and counted the members of more than nine thousand households quarterly). They interviewed families daily and collected data, asking practical, intimate questions: How many times did you breastfeed your baby today? Did you wash your hands after using the latrine? Did you use toilet paper? They wanted to know, What foods did you eat yesterday? Has anyone in your family fallen ill? What medicine did they take? They recorded this information on surveillance forms for Mal-ED. No one ever said the word *microbiome*.[64] Bacteria were only discussed in terms of disease, to teach handwashing and hygiene—ideas of beneficial or "good" bacteria were thought to confuse the issue, and as one icddr,b staff member commented during a presentation, "we aren't interested in commensals. We are trying to fight infection."

Clearly Gordon Lab scientists would disagree with this point, considering that all their work is predicated on the idea that entire communities of microbes matter, in both illness and health.

Three FRAs, Roji, Sunita, and Zarat, each introduced me to three of the families to which they were assigned. I spent eight hours a day, seven days a week with those mothers in hot tin rooms, sitting on pallets and beds, watching food being cooked, babies being nursed. These women ranged in age from eighteen to thirty, had one or two children, and most were born and raised in Mirpur. In *The Economization of Life,* Murphy has discussed the "Invest in a Girl" global campaigns in places like Bangladesh that turned questions of survival instead into modes of capital. This is reflected in pervasive statistics that boast Bangladesh as ahead of India in terms of girls' primary and secondary education, maternal health, and mortality. These neoliberal programs also led to "women-friendly" legislation that actually had the effect of deradicalizing feminist politics in the country—as Seuty Sabur argues, activists were faced with the untenable power of NGOs and the United Nations rather than of local governments.[65] The generation of women whom I met in Mirpur were embedded solidly within these programs and systems. Before they became pregnant, almost all were garment workers in Bangladesh's monstrous textile and clothing industry. Anthropologist Lamia Karim has written extensively about gender and labor in Bangladesh, especially about the establishment of microcredit programs for women. Though working conditions in the Bangladeshi garment industry are dangerous and wages are exploitatively low, Karim has shown that women who work in that industry are more socially and economically empowered than rural women participating in microfinance programs, because the former work outside the home and attain greater autonomy and self-awareness.[66] It was useful to see the Mirpur mothers in this context, in evaluating the unique population they represented. All the mothers in the Microbiome Discovery Project had extensive experience with various NGOs operating in Bangladesh, especially those focused on health education, hygiene, nutrition, and breastfeeding. The scientists studying the microbiome and I both operated in this context, capitalizing on icddr,b's relationship with the community, enrolling women who were familiar with foreign instruction about how to live their lives, were tolerant of personal questions from white people, and were willing to volunteer biological samples.

Maliha, mother of seventeen-month-old daughter Prianka, worked as an embroiderer, and her husband owned a small garment factory. Maliha grew up in the house we sat in; she lived there with her mother, husband, and two children.

> When I was a child, food was very tasty and fresh. Things like spinach don't taste good anymore, so my children eat less. These days people use chemicals on plants, like fertilizer, pesticides, and it makes food taste different. Soil has become very dirty and contaminated. In the past there were small amounts of plastic, but now there are huge amounts of plastic contaminating the soil that food is growing in.

Like other mothers in the study, Maliha was sensitive to changes in Mirpur since her childhood. For her, chemical exposures and plastic pollution had changed the quality and taste of food and caused her children's malnourishment. For this reason, instead of food grown in the contaminated ground, she often relied on crisps and cookies for snacks, bought at ubiquitous food stalls around the neighborhood. This evokes a nearly identical experience Emily Yates-Doerr described during her fieldwork with mothers in Guatemala: "this logic of health does not encompass the fact that women regularly purchased packaged chips and candies for their children when they were out shopping because they were afraid of microbes and pesticides."[67] FRA Roji confirmed this same practice for me: "Most of the mothers are so busy, they take outside food. Cake, ice cream, et cetera. Cohort mothers are much more conscious, they make all the supplementary food at home to avoid outside food." It seems maternal calculations of the competing harms of malnutrition and exposure can be seen in very different global settings.

Tea shops or "hotels" in the neighborhood sell packaged snacks like crisps and cookies.

Though gas and electricity services were now readily available, Maliha lamented that there were no open spaces left for children to play in. Packaged food was sold everywhere in the neighborhood and was expensive. Vanessa Watts (Mohawk and Anishinaabe) and Sandra Styres (Mohawk) explain Place-Thought[68] and Land[69] frameworks that connect humans and nonhumans relationally across time and space: "Place-Thought is based upon the premise that land is alive and thinking and that humans and non-humans derive agency through the extensions of these thoughts."[70] Shadaan and Murphy use these Land/body concepts while talking about exposures to endocrine-disrupting chemicals as colonial environmental violence "to indicate an understanding of land that is not commensurate with territory or earth, but rather includes nonhumans, ancestors, future generations, and 'all our relations' stretching both backward and forward in time."[71] These authors refer to Anishinaabek and Haudenosaunee teachings but acknowledge that ideas about Land generationality and capacious kinships are specific to many Indigenous nations. In examples like these, there is potential to use Indigenous scholarship when analyzing microbial enviro-social relations with humans. There is a connection here between Land/body relations and how Maliha describes her own childhood connection (and her children's disconnection) to the soil, to plants, to food. Generations of urban migration to Dhaka from rural Bangladesh because of food and job scarcity caused by colonial violence have contributed to extreme population density. Gut microbes are also knotted up in this, as they are inherited through familial practices and change generationally.

As mentioned, the icddr,b field researchers taught the study mothers the word *jibanu* when they were enrolling them in the study and as part of the hygiene and nutrition training. *Jibanu* is not a word in the everyday Mirpur lexicon, but mothers were told that it was *jibanu* that made their kids sick. For them, it became a concept bridging food, water, and places in the home to babies with diarrhea. Maliha takes seriously the microbial education the FRAs give her, and her responsibility as a mother to protect her family from microbes:

> *Jibanu* is in the latrine, dust, soil, rubbish. It is like little white worms that you can't see. It can be in food when you purchase it from the market. Sometimes I can see it and sometimes I can't—it comes in many shapes and sizes. It can make you sick. When we purchase the food and clean it very well the *jibanu* is not there anymore and when I wash the floor the *jibanu* is not there. But in the latrine, the *jibanu* is always there.

This is an important point, as mothers were encouraged to think in new ways about bodies and places: their hands; their babies' mouths; and how they touched pot handles, latrine buckets, and okra stalks. The icddr,b instruction taught that the external environment was both illness causing and invasive. Ultimately, these women began to think about how the bodies of *jibanu* circulated between the inside of their own bodies, the outside world, and back again. The problem of determinants emerges again, as Yates-Doerr explains, scaling problems both too broadly ("clean water," "sanitation," and "pathogens" disconnect enteric disease and malnutrition from the actual bodies and lives of individuals who are suffering) and too narrowly (public infrastructure, governmental corruption, gender violence, and religious racism aren't acknowledged as determinants because they can't be feasibly fixed). TallBear's intercession is useful here too, as she talks about "new" concepts of interspecies relations in anthropology and STS: "A second contribution to this growing subfield that Aboriginal thinkers can make is to extend the range of nonhuman beings with which we can be in relation. . . . But for many Indigenous peoples, our nonhuman others may not be understood in even critical western frameworks as living."[72] In Mirpur, we can see how the pot handles and okra stalks are part of food histories and generational knowledge and can become incorporated into systems of relations. One of the first things I learned about most families in Mirpur was that "households" were determined by shared cooking pots. Many people might share a stove or even a living space, but pots make kin.

On another day in Mirpur, I sit on a plastic chair that was maneuvered carefully from a neighbor's house, through two-foot-wide alleys and over the heads of the dozens of children crowding outside Mita's house. The flock of children are there because I am here, an attraction on an otherwise ordinary April day. Mita is a mother in the study; she is telling me that she knows about breastfeeding—she knows about doing it exclusively for six months, and she knows she is supposed to continue as long as she can. She holds sixteen-month-old Pavi on her lap and nurses her the entire hour that we are speaking to emphasize her point. But Pavi's status in the study has her diagnosed as malnourished, giving her a −3 Z-score.[73] Analyses of Pavi's microbiome say that the intervening breastfeeding education has come too late. Now the baby must be fed therapeutic foods to make her healthy, and the scientists running the study hope someday her broken microbiota will be repaired with prescribed personalized probiotics. The Gordon Lab's second-phase "Breast Milk, Microbes, and Immunity" study is using samples like Mita's and Pavi's to determine what comprises "normal" breast milk within and across various human populations to find out how nutritional recommendations for infants, children, and their mothers should be tailored to the state of functional maturation of their gut microbiomes. Later, the director of the Nutrition Programme at icddr,b will tell me that it has been hard to break urban Bangladeshis of their "cultural" habits. Mothers are compelled to feed babies prelactates (things they feed even before breast milk)—mostly sugar or honey meant to make the baby's disposition sweeter. The mothers themselves have poor nutritional status and for this reason often think their milk is insufficient or bad. They discard the supernutritive colostrum as "spoiled" and think a baby crying always means the baby is hungry, the baby must need supplemental foods. Adeela is the FRA supervisor for the Mirpur site. She has a master's degree in child development and family relations from Dhaka University and has been with icddr,b for six years. When I asked about breastfeeding in the study cohort, Adeela explained, "We also teach them about hygiene, most of the time cohort mothers also get breastfeeding counseling. Most of the time they complain their children aren't getting enough breast milk, Motia is a breastfeeding counselor. They realize that they *do* have enough milk, that it is only in their minds that they don't have enough milk for their children. Sometimes it is very difficult to breastfeed, position is very important. Breastfeeding is not easy." There is a hybridized form of knowledge circulating where generational practices are confused with NGO mandates and compliance to these mandates is tied to health care, resources, and services. Mothers are reprimanded for following traditional practices that are viewed as harmful—yet ideas that their own bodies are insufficient or that their first milk is "bad" are reinforced by relentless public health narratives about nutrition and health.

The lability of environment in microbiome science makes flexible concepts of bodies, kin, exposures, and sociality—and what value there is for different actors to concretize different meanings. Study mothers focus on pathogenic microbes, think about generations of family, define environments as house, market, bodies, food. The Gordon Lab focuses on microbes as evolutionary kin, defines environments as genetic, intestinal, geographic. As Adeela tells me:

> when I see a malnourished child, now I think, what is going on with that child, what are the factors in his life that are causing this? The scientists have to come here to see it. There aren't little things I can express, unless they come here, they don't know what it's like. I can only tell you my opinion, and it is too hard to explain the conditions here without seeing it for yourself.

Studies of the microbiome in Dhaka are studies of bodies merged with exposures and embedded in kinships. Adeela doesn't just mean that the state of things in Mirpur must been seen to be believed; she is saying that both human and microbial bodies are materially affected by the social circumstances of everyday life in Mirpur. She knows that the icddr,b and Gordon Lab scientists receive hundreds of thousands of biological samples—she and her team collect, label, and pack them. She also knows what prescriptives about breastfeeding and antibiotic use circulate back to Dhaka from the United States after samples are analyzed. Adeela mentions to me more than once that these mothers don't care about microbes, that getting free health care for their children, keeping them fed and without diarrhea, is all that matters. Microbes act and are enacted differently between worlds, and local microbiologies[74] become significant in different contexts for people who have divergent stakes in their respective relationships with microbes.

While microbiome research promises eventually to treat biocentric malnutrition by fixing a "broken" microbiome,[75] locating microbes in breast milk, food, and homes adds weight to the already heavy load of maternal care and duty, especially for pregnant mothers.[76] Bacterial DNA found in amniotic fluid, umbilical cord blood, and placentas led some researchers to speculate that human microbiomes are established before birth, passed from mothers to fetuses during pregnancy.[77] The majority of others argue that the presence of placental or meconium bacterial DNA only signals sloppy science and contaminated samples.[78] Some point out that decades of breeding gnotobiotic mice with sterile uteruses that gestate germ-free babies prove definitively that there are no microbes in the placenta.[79] As of this writing, the debate rages on.

The environment's multiple and microbial kin split and converge again. In this instance, the environment becomes the prenatal world of uterus and placenta, and microbial kinships are transmitted before birth. Mansfield and Guthman have discussed how, in an epigenetic age, "the 'uterine environment' is figured as the key space-time of epigenetic becoming."[80] If the gestational environment can affect the genetic and disease outcome of the fetus, the mother is responsible for the maintenance of this environment. If the microbiome is seen to be established in utero, "the womb is the environment that must be controlled because it is the fetus through which the past becomes the future."[81]

Lappé and Jeffries Hein continue a similar critical analysis about the increasing significance placed on placentas in relation to developmental origins of health and disease: "we described how placentas are imagined and enacted as not only a connection between bodies, but as a means of understanding relationships between lives, environments, and time."[82] The human microbiome has refocused "mother-to-child transmission" on the transfer of commensal and beneficial bacteria from birthing parent to baby, concluding that important host–microbe partnerships are actually heritable. "Traditional," premedicalized activities like vaginal birth and breastfeeding loop back around as important to healthy microbiomes and scientifically condoned best practices. Yet the profound responsibility of mothers remains, shifting from prevention of harm (pathogens) to establishment of good (microbiome).

In the connection between birthing parents, environment, and microbes, bodies have permeable edges; organisms exchange genes, coevolve, and influence development across generations. Consider this quotation from the gastroenterology and hepatology journal *Gut*:

> Infants are naturally born with their skin and mouth covered by maternal inocula and have swallowed these microbes supported by the observation of both DNA and live bacteria in the meconium. Thus, we *inherit* the primordial microbiota from our mothers, grandmothers and further on the matrilineal line, with microbial vertical transmission extending back to earlier ancestors.[83]

Human microbial ecologists call this the "primordial" or "ancestral" microbiome, and in reference to this, I once heard a scientist say "the vagina is the center of the universe." It is important to account for what it means for us to share kinship with microbes, not just as biological kin. If we are intergenerationally, socially, and environmentally linked to microbes, we must also think about what politics, violences, and capitalisms those universe-centering vaginas were also a part of.

Because nearly all that is known about gut microbiota is gleaned from the sequencing of bacterial genomes found in fecal samples, to know microbes is to rely on the future of the newest genomic sequencing techniques. However, human–microbe relationships themselves are ancient and ancestral. I want to stay with kinship. There is an interesting coincidence in this moment, when we as anthropologists are looking closer at multi-/interspeciality and accounting for the material in new ways, yet also seeking a credible voice within the biological sciences. There is a vital need to reframe these categories and do transborder theorizing to see nonhuman relationships not just as things that are good to think with but as things that might be crucially necessary to alleviating human suffering. I advocate for a persisting use of kinship as a concept, especially when it comes to microbiomes, agreeing with Stefan Helmreich when he says that new facts of biogenetic relationships should be viewed as kinship facts, "because so seeing them allows for the deployment of other kinship tropes of shared responsibility and risk, solidarity and dispossession."[84] Kinship for anthropology has meant a set of social relations, various, complex connections, but it could also mean a framework through which we advocate for justice. Nonhuman relations are intimately enmeshed in our life-or-death stakes. We cannot have coevolved for nothing.

As we parse what social determinants of health means in all its indeterminant parts, and what work it does or fails to do,[85] what we mean by *environments* and *kinship* may be refigured through a focus on microbes. An anthropological intercession can add ethnographic data corresponding to biological samples, scrutinizing the place and the science that bring microbiomes into being. Observations on ways of living, caring for children, and feeding families can help broaden the scientific view of study subjects. A medical anthropology and human microbial ecology collaboration can account for the sociomaterial conditions of everyday life and the vulnerability of all kinds of exposures in studies of the microbiome. Considering Indigenous theories on kinships and environments across generations, we can rethink our relationships with microbes and our relationships to one another through microbes.

During my many years of transdisciplinary partnership with microbiome scientists, I have been asked, what would microbiome research look like if it could actually integrate, acknowledge, and learn from Indigenous and postcolonial scholars and their own study participants? I don't really know, because so far, an anthropologically engaged microbiome science fails more than it succeeds. This partial truths orientation creates possibilities to rethink rather than return to the human, disrupting the politics of care: Which bodies and which environments do we account for? Where do we distribute medicine, intervention, and justice? This experimentalism in multiple registers shifts the boundaries of environment, relations, and self for scientists and for anthropologists—making microbes into kin.

A large bowl of chiles in Mirpur after they have been dried in the sun.

FOUR

Malnutrition Futures

Group Therapy

I'm sitting in the Gordon Lab conference room attending Group Therapy, a weekly Tuesday, 9:00 A.M. meeting in which lab members (minus Dr. Gordon) discuss the practicalities of being a student or postdoc; describe their works in progress; and, in journal-club style, review recent microbiome publications. The lab manager tells me Group Therapy used to be a forum for airing grievances; it used to be less formal, and people would actually talk about problems they were having with their work or in the lab. Now lab members use email for communicating those things—digital social networks have changed the way lab relations (personal and professional) get done. As the meeting starts, someone says, "Hey, is there a microscope around here anywhere?" Michelle, a postdoc, answers, "There's one in the back room by the PIs' offices—it's a nice 'scope, but no one knows how to use it." As weird as it seemed to me, a lab wholly dedicated to the study of microbial life had no instruments for "viewing" microbes with the eye. In 2010, the practice of *seeing* microbes through genomic and metabolomic technologies was rather nascent, unchartered territory. This chapter addresses three concerns: how the microbiome is produced as an experimental object in the context of metagenomic knowledge, what social science approaches to microbiome science have subsequently emerged, and how knowledge about microbiota (and nutritional states) is crucially bound to the highly specific technologies used to sequence microbial genomes.

Conditional Data

One-third of childhood deaths globally are linked to poor nutrition, and in what global public health calls the "developing world," one-third of all children are malnourished—suffering dire physical and cognitive consequences. The designation of "developing" is given by the World Bank, in relation to gross national income per capita, to nations that have not achieved significant industrialization or technological infrastructure relative to their populations. An anthropology of development has a contrary analysis, focusing on the specter of exploitation, colonialism, and racism that overshadows categories of "developed."[1] The worldwide management of malnutrition has been deemed a failure by the WHO and World Bank, and major global health funders, such as the Gates Foundation, have prioritized the creation of new tools to contend with undernutrition. Malnutrition is the biomedical category of poor health conditions caused by an imbalance between the nutrients a body gets and what it needs, which can be caused by a lack of food, calories, or micronutrients (undernutrition) or an excess of calories or nonnutritive food (overnutrition). While conventional attempts to address hunger and protect food security endure, new approaches to malnutrition in the life sciences have turned toward the use of big data techniques to analyze the gut microbiota of children suffering from these devastating problems. Using metagenomic technologies to evaluate the metabolic function of microbes in the gut, scientists seek causes of malnutrition at the interface with the microbial environment, research that is expected to manifest in malnutrition solutions for humans based on their microbes.

I followed gut microbes as they traveled—physically and conceptually—between the Gordon Lab, using human microbiota in diet/genomic studies modeled in germ-free mice and the field: the childhood nutritional intervention study in Bangladesh. The mothers I met in Mirpur collected daily fecal samples from their babies that would eventually produce shotgun-sequenced microbial genomes and 16S rRNA reads that transformed microbes into new forms of scientific knowledge. This chapter proposes a doubled focus: first, on the social and material conditions of the Bangladeshi women and children enrolled in the microbiome study, and second, on the *datafication* of microbes that facilitated drawing a causal relationship between microbial populations and undernutrition in their human hosts. Datafication has been used to describe the processes of turning life into data and data into new forms of value. Here I use datafication to talk about the ways in which microbial behavior is analyzed through data-driven science and how datafied microbes link back to the lives and health of humans. How is malnutrition lived for urban Bangladeshis, and how is malnutrition studied through metagenomics? What categories of undernutrition and bodily health emerge as big data becomes a tool for nutrition science? The Gordon Lab is at the forefront of studying how human microbiota is constituted and changed and its role in health and disease. The lab's research aims to develop a translational pipeline for identifying next-generation prebiotics, probiotics, and synbiotics to prevent or treat metabolic dysfunction and nutritional deficiencies.

My time observing and working in the Gordon Lab was interspersed with trips to Dhaka to conduct fieldwork investigating social determinants that affect microbiota and malnutrition. I provided the lab with qualitative ethnographic data about living conditions and social networks in Mirpur. In addition to scientists in the Gordon Lab, and those at icddr,b, I spoke to experts in the wider fields of gnotobiotics, microbiology, molecular genetics, and public health. Ethnography as an analytical approach is a critical first- and second-order methodology in this work. As discussed, I worked in the Gordon Lab, studying and learning from scientists and their practices, and I also was a member of the scientific team and a contributor to the work, forging what I hoped would be a "post-ELSI"[2] "friendship with scientists."[3] Because commensal microbes are transmitted and cultivated between humans through biological and social intimacies, human microbial ecology and anthropology became co-accountable to contextualizing and collecting data for the research project in Bangladesh.

At the Gordon Lab collaborative field site in Dhaka, I interviewed Bangladeshi FRAs and women enrolled in the microbiome maternal–child malnutrition project. Mothers and I talked about food, home, and history, while samples of blood, urine, and baby feces were collected by FRAs. Back in the lab, germ-free mice were humanized with microbes taken from these fecal samples and given diets designed to be representative of Bangladeshi food, their guts then sectioned and sequenced. The present microbial communities were analyzed using high-throughput, culture-independent metagenomic methods. Lab members used these results to learn how many physiologic, metabolic, or immune features of the human donors could be transmitted to recipient mice, and using this information, they began to establish reciprocal relationships between human gut microbes and health status. Microbiota from inside people was manipulated in the guts of mice and translated into large sets of genomic data used to understand the structure of disease.

Mayer-Schoenberger and Cukier coined the term *datafication*,[4] and it was popularized for the social sciences by Van Dijck,[5] who used the term to describe the transformation of social behavior into quantified data for mining, analysis, and prediction. Big data's ubiquity in socioeconomic arenas has led many social scientists to argue for a more synergistic view of ethnography and big data,[6] has led to the formation of the NSF-funded Council for Big Data, Ethics, and Society and the creation of critical research institutes like Data Society. In 2015, *Big Data and Society* was created—a peer-reviewed, scholarly journal connecting debates about the emerging field of big data practices and how they are reconfiguring social, scholarly, industry, and scientific realms. Clearly the potential impacts of big data on human lives are on the social science mind.

My analysis starts with the recognition that no data are value-free: "just because we have big (or very big, or massive) data does not mean that our databases are not theoretically structured in ways that enable certain perspectives and disable others."[7]

Furthermore, as Leonelli has deftly shown in her analysis of the effects of big data on the biological sciences, data's value is not in their uncomplicated depiction of phenomena but in their portability and potential as future evidence. The methods, tools, and knowledge of data handling and analysis influence how scientists conduct research and, consequently, how they understand biology.[8] In this chapter, I am using datafication to examine the ways in which microbial functions are datafied through metagenomics and how, subsequently, microbiome data are connected to the diets, health, and lives of humans. Using an ethnographic lens to analyze different enactments of malnutrition allows for an exploration of different perspectives on data: Bangladeshi families' multiple, multifarious experiences with stunting, poor eating, and diarrheal disease; the context and content of nutritional interventions; and how implicit assumptions are made about the correlation between 16S rRNA data sets from fecal microbiota and malnourished bodies.

This chapter is organized into four sections. First, "Mirpur Mothers" looks at practices in homes and communities in Dhaka and how these practices intersect and conflict with the biomedical and public health diagnoses of "malnourished." "The Sack" thinks through microbial scaling; approaching our microbiota as a metabolic organ that crucially affects human health is a big-picture, big data concept in nonreductionist, postgenomic science, yet the practice of studying microbes requires dealing in very small substances, small creatures, and nuanced scientific maneuvers. Third, "Computational Malnutrition" looks at how malnutrition is enacted in the lab and how it is choreographed with the data-producing technologies of microbiome studies. Last, "Datafuturity" looks to the implications of this analysis: what the direction of research means for scholars of different disciplines; as gut microbial communities are datafied with the aim of global nutritional intervention for humans, what vectors of health and illness are set into motion. Oscillating between these two field sites—Bangladesh and the Gordon Lab—being an ethnographer of and for scientific microbiome research, I attempt to collaborate with scientists to develop more holistic analytical and methodological frameworks, with the goal of understanding the embodiments of malnutrition within dynamics of suffering and poverty. Operationalizing ethnographic knowledge, and facing what corresponding compromises[9] and uncertainties[10] result, is an ongoing challenge of this work.

Mirpur Mothers

Kanta does not like eating. She is eighteen months old and quick on bare feet, out the open door of the nutrition center into the hot Dhaka street three times before her mother, Rimi, catches her. Typically gentle and affectionate with her daughter, Rimi holds Kanta firmly in the green light of the nutrition center, under illustrated posters teaching about nutritious food and handwashing. Kanta cries loudly as Rimi tries to spoon cereal into her mouth. "She hates the *pushti* packets, but she usually doesn't like any food, so we force her to eat."[11]

RUTFs are nutritional, dietary supplements, primarily used for emergency feeding of malnourished children. The foods are referred to as lipid-based nutrient supplements or milk-based fortified spreads. The aim of RUTFs (for crisis feeding, or ready-to-use supplemental food, RUSF, for chronic undernutrition) is to have a prepackaged food that can be used at home without medical supervision; that resists bacterial contamination; that has a long shelf life; and that requires no water, preparation, or refrigeration. The ease of transport and administration of RUTFs allows for low-cost, mass treatment of malnutrition. The first patented and most widely used RUTF was Plumpy'Nut, a peanut paste mixed with sugar, vegetable fat, and milk powder, enriched with vitamins and minerals. Plumpy'Nut is manufactured by the French company Nutriset and has been used extensively in African famine situations. While Plumpy'Nut has been heralded as a miracle cure for global hunger, complications and controversies have arisen. Nutriset has been accused of aggressively protecting its patented product and of preventing other organizations from providing lifesaving RUTFs to starving children. As a result, Nutriset developed PlumpyField, a network of local Plumpy'Nut producers in countries affected by malnutrition in an attempt

Mothers were shown the ingredients for the *pushti* packets so that they knew what their children were eating and felt more comfortable with foods they recognized.

to cultivate "nutritional autonomy."[12] And though the biological health benefits of RUTFs have been well established, little has been done to examine the economic, political, social, or even gustatory implications of therapeutic food. One Russian graduate student in the Gordon Lab commented that she thought RUTFs were "American-centric," that it is a Western assumption that all children are going to love the taste and texture of peanut butter. One importantly critical voice in the use of RUTFs is public health expert Lora Iannotti, who is interested in the complications as well as the benefits of RUTFs. She looks at taste, texture, and eating practices, and her work in Haiti exposes the informal economies that spring from nutritional intervention—the selling, trading, and sharing of RUTFs.[13] How gut microbes react to the introduction (and cessation) of RUTFs and how children and families incorporate or reject these foods are examples of microbial scaling. RUTFs create specific practices, meanings, and biologies of eating that influence and are influenced by microbial populations. Much better luck has been had by icddr,b producing therapeutic foods made from local ingredients mothers are familiar with, and ones that families like to eat. Along with *pushti* packets, which are provided at the nutrition centers, FRAs also instruct the mothers on how to prepare *khichuri,* which is made of greens, squash, potatoes, lentils, and soybean oil.

The Bangla word *pushti* translates to "nutrition" in English. Although half of all infants in Bangladesh weigh less than 2.5 kilograms (5.5 pounds) at birth (almost 1 kilogram, or 2.2 pounds, under WHO birth weight standards), it is when children get very sick that mothers seek help from icddr,b. Nationally, 41 percent of Bangladeshi children under five years of age are moderately to severely underweight, and 43.2 percent suffer from moderate to severe stunting, an indicator for chronic malnutrition.[14] *Kwashiorkor* is the most common form of severe acute childhood malnutrition diagnosed by icddr,b; it presents abruptly, usually with edema (swelling of ankles, feet, and abdomen), and causes physical and cognitive stunting.

The word *kwashiorkor* is derived from the Ga language of coastal Ghana and means literally "the sickness the baby gets when the new baby comes," as the first child is "deposed" from the breast when a new child arrives. It was implemented in global medical discourse in 1935 by Dr. Cicely Williams, a white Jamaican who worked on maternal and child health for decades in Ghana and Malaysia. She believed it to mean that a first child was displaced, weaned, and consequently became undernourished once the second baby came.[15] During my fieldwork, *kwashiorkor* (colloquially, *kwash*) was the word used commonly around the Gordon Lab, in Dhaka, and in published papers and was the primary type of malnutrition diagnosed in the Microbiome Discovery Project in Bangladesh. In subsequent years and studies, *kwashiorkor* has been replaced by *severe acute malnutrition,* or SAM. Malnutrition itself as a complexly historical and colonial category cannot be taken for granted. History of science, technology, and medicine scholar Michael Worboys makes the claim that "third world" malnutrition was "discovered" in the then-colonies between the World Wars, "constructed as an imperial problem and put on the world political agenda."[16] The health (or unhealthiness) of a population became a matter of nations and determined by mapping a colonial metric onto a colonized population. Malnutrition came to be a problem of the "underdeveloped" world, a problem that, with the emergence of a global health framework, transcended national borders and became biologically universalized, solvable through neoliberal economic strategies, and measurable through standardized metrics.[17] Malnutrition continues to manifest in particular places in the world, at certain times that correspond with global politics, economies, and moments in science and technology. As Landecker points out in her "biology of history" framework, "we come to inhabit the material future produced by what we thought we knew."[18] Rarely are the past colonial exploits of land and people considered in evaluations of modern malnutrition.

Malnutrition status for children in Mirpur is determined by the health care workers, first by sight, based on their extensive experience, then by measuring bodies. Severe or moderate acute malnutrition translates to a −3 or −2 Z-score. The Z-score classification is the standardized biomedical system used to measure malnutrition, interpreting weight-for-height, height-for-age, and weight-for-age. The Z-score system expresses the anthropometric value as a number of standard deviations or Z-scores below or above the median value. The WHO advises that Z-scores are not generalizable across geography or living conditions and should be used only as a general guide for screening and monitoring and has done work in the last ten years to better adapt its metrics across different kinds of bodies, looking to environmental conditions as the greatest factor in growth metrics.[19] The previous, much reviled National Center for Health Statistics/WHO growth reference recommended for international use since the late 1970s was based on a longitudinal study of children of European ancestry from a single community in the United States, clearly inadequate for children worldwide in different socioeconomic, nutritional, and environmental conditions. The currently recommended Multicentre Growth Reference Study (MGRS) standard gathered primary growth data from 8,440 healthy breastfed children from widely diverse ethnic backgrounds and cultural settings (Brazil, Ghana, India, Norway, Oman, and the United States). For scientists working within the malnutrition study at icddr,b, it follows that children in any population can be compared with the MGRS growth standards, allowing for straightforward comparison of growth metrics across populations, with growth being much more dependent on individual environmental growth factors than ethnicity.[20] Mothers in Mirpur don't have a definitive definition of "malnourished"; it is a medicalized metric that is given to them, told to them. In their experiences, there are much blurrier lines between healthy and sick, growing and stunted.

Important work has recently been done in the social sciences on the "double burden" of malnutrition in "Westernizing" countries, where new foods lead to an overconsumption of fats, proteins, and carbohydrates, while people still underconsume sufficient vitamins and minerals. These works focus on "obese" bodies and the ethnographically substantiated, critical unpacking of the designation of "obesity."[21] However, we have yet to see work that takes apart designations of undernutrition and undernourished bodies in the same manner (nor do I do that here). To be sure, those from Malthus to Foucault to Scheper-Hughes have thought about the social/economic/political significance of the "starving body," but without questioning the material reality of such a body. Clearly overnutrition and undernutrition are not two equal sides of the malnutrition coin.

Gordon Lab members often pointed out that what would be considered "malnourished" in St. Louis is not "malnourished" in Dhaka. How malnutrition is lived in can also be situationally different; urban Bangladeshis don't face food security crises resulting from scarce harvests or drought but instead are confronted with chronic enteric disease from the beginning of life. Dr. Gordon often came back to this perspective: "There are many continuous and categorical, potentially interacting explanatory variables affecting malnutrition: health status, diet, antibiotics, infectious diseases, gender, age, region, family. Relationships between these variables aren't linear. Some monozygotic twins in our studies are allowing to control for genotypes, but could some elements of malnutrition be epigenetic?" Within Gordon Lab microbiome studies, malnutrition is seen not as a wholly biological condition, but as one that is the result of integrated inter- and intragenerational, genetic, socioeconomic, environmental, and microbial components.

After icddr,b health workers diagnosed Kanta with kwashiorkor, Rimi enrolled her in the microbiome nutrition study. "I thought my baby had a vitamin deficiency, and most importantly, I wanted medical care for her. It is very good for my baby, because she was underweight, and now she is in good health. I don't mind giving the feces and blood samples, they can use them to find if my baby has any diseases, they will know about it and can treat it." Kanta's fecal samples, collected by Rimi, are packed on ice in picnic coolers, travel to the icddr,b hospital, and are eventually flown back to the Gordon Lab. There they are inserted into the guts of germ-free mice, which are then fed a specially lab-made, representative "Bangladeshi diet," bringing together biological and social representations of life in Dhaka. Malnutrition is analyzed through the interconnected functioning of human and microbial genes read through big data–producing technologies, machine learning, and metagenomics. Bangladeshi children become "necessary subject-objects,"[22] their bodies and experiences the basis for the transnational research and its technologically imagined solutions to malnutrition. In chapter 3, dovetailing Lappé's reflections on "the gendered dimensions of care and the less often acknowledged forms of affective and material labor that support scientific research,"[23] I discuss the ethical complexities that result for Mirpur mothers and babies: the motivations and risks of enrollment; the weight that locating microbes in bodies, food, and homes adds to the already heavy cultural and biological load of maternal duty; and the power mothers have in providing both fleshy material and daily information to the scientific research.

Following Lock's concept of "local biologies," in which biological and social processes are inextricable and coproducing (resulting in material, biological differences in humans),[24] anthropologist Erin Koch has devised "local microbiologies," placing a focus on human–microbe relations, with the idea that "microbes are part of unique social matrices, and their biochemical attributes are only made meaningful in particular contexts."[25] I see local microbiologies in action within microbiome–malnutrition studies, as certain microbial attributes become meaningful in different contexts for people who have a stake in their respective relationships with microbes. For Rimi, malnutrition comes into being through concern for her daughter, days of diarrhea, and the things she must now perform as an enrollee in the study. Rimi learned about malnutrition (*apushti* in Bangla) and bacteria (*jibanu,* literally "causative organism") from the FRA who enrolled her in the study; before this, she knew that Kanta was very small at birth and that she was often sick. Her relationship with microbes becomes represented by relentless handwashing; new cooking and living practices; daily walks to the nutrition center; and interactions with scales, syringes, and plastic tubes. She understands *jibanu* to be related to Kanta's unhealthiness, and tells me:

> *Jibanu* is in the water, it is killed when water is boiled. They look like little insects but are invisible. In fish hatcheries, they put a lot of chemicals in the fish. *Jibanu* is in the fish. Also in vegetables. It can get in your body if you eat infected food, and you will get diarrhea. *Jibanu* can be in the latrine and in exposed food. It affects everyone the same, young or old.

Rimi knows she cannot see *jibanu* but thinks of it as a shadowy threat in her family's water, food, and feces; illness from *jibanu* is interconnected with ongoing water crises in Dhaka, unnatural additives to farmed fish, and the sanitary conditions of homes. The local microbiology at work in Mirpur is one of responsibility and diligence, new conceptualizations of food and environment, and learning about how *jibanu* will interact with bodies. The Gordon Lab is focused on microbial genomics and, at the same time, concerned with the environments[26] (diets, homes, lifestyles) that produce the microbial populations—how things outside the body are converted into the biology of the body.

For Gordon Lab scientists, big data–producing technology enacts malnutrition within the microbiome, and local microbiologies in the lab focus on the metabolic and genetic tasks microbes are performing. This genomic functionality—whether the microbes have genes for vitamin biosynthesis, are poor at breaking down amino acids, or are loaded with genes for breaking down complex sugars—becomes meaningful when looking for the distinct relationship between organisms in a human gut presumed to be healthy and those in a gut that has been marked as malnourished. As scientists use bioinformatic means to identify microbiota, the microbes themselves become datafied. For example, to "see who's there" in the microbiomes of healthy and kwashiorkor-diagnosed (using WHO criteria) twins, metagenomic analyses use DNA prepared from the children's fecal samples, subject it to multiplex shotgun pyrosequencing, and compare the reads with databases of already-known microbial genomes. The Gordon Lab used big data techniques like principal coordinates analysis to explain the amounts of variation between the databases and to assess functional development of the microbiomes of twin pairs who remained healthy and twin pairs who became discordant for kwashiorkor.[27] Simply, differences in malnourished and healthy microbial populations are culled from datafied information about samples from phenotypically malnourished and healthy donors. Thus local microbiologies drawn from Mirpur become big data through Gordon Lab analysis, in order to map treatment back onto local Bangladeshi biologies.

Kanta lives in Block D of Mirpur 11, in a three-room house with tin walls. In one room, Kanta sleeps with her mother and father; in another, her grandmother and grandfather sleep with their nine-year-old son. The third room is occupied by a renter and her young daughter. Eight people inhabit this twenty-by-twenty-four-foot space, sharing the attached bathroom and narrow kitchen, a supply water pump and pit latrine, two gas burners. In Kanta's house, there are no doors, no windows, no shoes. On her street, there are chickens and ducks underfoot and an open drain in front of every house. Rimi cooks food for the whole day before the gas service goes off at 9:00 A.M., preparing huge pots of rice, dal, and spinach and potato mash and negotiating time at the stove with the five other families who share it.

The way in which Kanta's microbiome comes into being is dependent on its social and biological locations. Along with birth, breastfeeding, and genes, it is shaped by local practices surrounding water, chickens, and bare feet. The microbial populations in Kanta's gut are specific to the dilemmas that these bring into being. But how to account for these social and environmental factors? Although food scarcity, hunger, and enteric disease have always complicated human lives, malnutrition is a complexly historical, context-dependent, sociobiological state, in this case, informed by microbes. Gordon Lab members were mindful of the many potentially interacting explanatory variables affecting malnutrition—health status, diet, antibiotics, infectious diseases, gender, age, region, family—yet correlating these conditions to data or integrating them into analysis is an ongoing problem.

In the introduction to a special issue of *BioSocieties* on "Alimentary Uncertainties," the guest editors employ the term *aliment* to interrogate what is scientifically knowable about food and health, highlighting that what nourishes the body extends beyond what is eaten.[28] Alimentary thinking is useful to me here, especially in considering "how incommensurate data sets, metabolic processes, models of eating behavior and public health interventions are experimented with and adjusted, with the output of policy in mind."[29] The daily home, cooking, and personal hygiene practices of families in Mirpur intersect, conflict, and are shaped by the biomedical diagnoses of "malnourished." Most urban Bangladeshis face issues of food accessibility and affordability rather than food availability[30] and are confronted with abysmal water/sewage conditions and enteric disease from the beginning of life. One field doctor in Dhaka told me he estimated that children who suffer from the most severe chronic illnesses can experience up to three hundred diarrheal incidents before age three. Mirpur mothers facing these conditions are incorporated into the study and encouraged to "know themselves biomedically"[31] by adjusting their interactions with the environment: wearing shoes to the latrine; using soap; washing hands after defecating; washing vegetables; feeding babies with spoons instead of hands; and, perhaps in the future, ingesting prescribed, personalized microbial therapies. Microbiome scientists hope to use computationally derived data to develop probiotics to circumvent sociomaterial vulnerabilities, microbially and preventatively.

The Sack

Gordon Lab members affectionately call their gnotobiotic facility the "Mouse House." It is the largest of its kind in the country, housing thousands of germ-free animals (as well as those with defined, controlled populations of bacteria), and scrupulously maintained by the Gordon Lab team. The Animal Facilities building is across the Washington University Medical School campus from the actual lab itself. It's an industrial-looking, unmanned building requiring the swipe of a special ID card for entry. This is where the animals are, where the flesh of the experiments lives, where the dirty science is done. Mice are another small thing in this story, yet they are producers of big data. Studies of microbiota depend heavily on mouse models—the standard experimental protocol is to transplant the microbes from a human fecal sample into a mouse gut.

Gnotobiotic models incorporating the gut communities of individuals representing different ages, physiologic phenotypes, and lifestyles could help forecast the nutritional value of foods currently being produced or anticipated for geographically and culturally distinct consumer populations, inform the design and interpretation of clinical studies, and help industries and governments as they consider how and what to feed people as our population grows to nine to ten billion by mid-century. To be successful, a "holistic" approach will be required that includes consumer education about the microbiota; an evolved vocabulary that meaningfully describes diet-by-microbiota interactions; a regulatory system prepared to process health claims; and integration across a number of disciplines, including anthropology.[32]

The literature concerning experimental animals, organisms used in scientific research, has become a strong subset of animal studies work.[33] I am taking the definition of animal studies broadly to be the field in which animals are thought about in a variety of disciplinary ways in an attempt to understand human–animal relations. These texts address the complex contradiction of accepting animals as human enough to be biological proxies in scientific work yet not human enough to render them unkillable. The laboratory animal body as transcendental, as incorporated into scientific practices and processes, as a collection of discrete parts, as a commodity in the economy of science, and even as the recipient of care and love is well discussed in literature.

Twenty yards from the Mouse House, there is the smell of animal urine, which only intensifies as you enter a large, two-sided elevator, knock on a locked, light blue steel door on the fourth floor, and enter a small office with a tiny desk and a cart with strange, slender canisters. A bulletin board above the desk displays pictures of the same blond child at different ages: the son of David O'Donnell and Maria Karlsson, the husband-and-wife team that codirects the gnotobiotic facility. They are mouse experts, nurturers, and killers. David, Maria, and the research lab manager, Sabrina Wagoner, do everything here. They breed the mice, feed them, clean their cages, and sacrifice them. *Sacrifice* is commonly used by Gordon Lab scientists and techs to describe methods for killing laboratory specimens.[34] David and Maria monitor everyone's experiments and determine how many mice (and what kind) and how many isolators are available—and if the experiment can be done at all. Though David and Sabrina vehemently deny they are scientists or are performing science (they consider themselves technicians, or facilitators of science, respectively), they are well versed in genetics, anatomy, and embryology. The mice guts are where one type of scientific data is produced, materially coming into being. It is where hypotheses and experimental design meet food, ovaries, and genes. Through one door, the small sacrifice room where I now stand. Through another door, the mice.

It's July in St. Louis; outside, it is sickeningly hot and stiflingly humid. For the sake of the mice, the Mouse House isn't kept as cold as other lab facilities, and the room smells of mouse chow, ethanol, and urine. I'm wedged between two gigantic liquid nitrogen tanks and a sink; in front of me is a lab bench with Styrofoam coolers with dry ice for snap-freezing tissue, a metal thermos, plastic cutting boards with rulers, Petri dishes, glass plates, prelabeled test tubes, little plastic filters that look like thimbles, foam-tipped swabs called lollipops, scissors, forceps, tongs, and aluminum foil. Present are two postdocs, a graduate student, the lab manager, and me. There are six people and six mouse cages in the room, and it's crowded. For what it's worth, I also happen to be newly pregnant and extremely nauseated.

The animals are twelve-week-old, black 6 mice, born to germ-free mothers, kept in germ-free isolators, and raised without any microbes until they were ten weeks old. In a process common to all Gordon Lab animal experiments, these mice were humanized with gut microbes from human donors. Gordon Lab members use this process to see what certain microbiota do under certain circumstances (e.g., how they respond to diet changes), and in germ-free mice, they can control both the microbial communities and the animals' diets and environments. Previous proof-of-principle experiments have shown that human phenotypes can be reliably and reproducibly captured in germ-free mice by human microbiota transplantation,[35] though as David tells me, "you never get out what you put in." A *proof-of-principle* experiment is one that proves the feasibility of a concept or method. Commonly known as black 6 (B6) mice, C57BL/6 is the most popular inbred strain of laboratory mouse and the most regularly used experimental animal model worldwide. The first mammal to have its genome sequenced, this strain is very controlled—each mouse is a genetically identical clone, scientifically engineered for research. Because the animals are genetically the same, experiments using them should be easily replicable. The genotypic consistency of these mice perpetuates their use in labs—the more research/scientific precedence that is based on work with B6, the more they are used. They are preferred as model organisms because, like some humans, they have a high susceptibility to diet-induced obesity, type 2 diabetes, and atherosclerosis. B6 mice are described as good breeders, robust, irritable, and "little alcoholics," because, unlike other mouse strains, they voluntarily drink alcohol.[36] As Karen Rader has written in *Making Mice,* her history of standardized experimental animals, the genetically homogenized mice became the overwhelming model because they were able to bridge the gap between nature and the scientifically created.[37]

The particular mice for this experiment were gavaged with the selected microbiota at ten weeks; a syringe containing the pulverized fecal sample is put down their throats, the plunger is pressed, and with it, an entire ecosystem explodes into their germ-free guts. Before gavaging was the norm in the Gordon Lab, microbial samples would be wiped on the mouse's fur so that the mouse would lick it off and orally ingest the microbes. Using this method caused uneven colonization in the gut (some bacterial strains would die off), so now germ-free mice are always gavaged with microbiota. In this case, the microbes have come from three set of human twins, all discordant for obesity (discordancy meaning that one twin is deemed obese and one is not, based on MRI scans of their body fat). As has been historically true in biomedical and psychological research,[38] twins are particularly interesting to scientists investigating the microbiome. Twins presumably serve as original genetic copies of one another, and it is thought that any health disparities between them reveal epigenetic and, in this case, microbial variables.

Like many of the experimental protocols for microbiome research, a fecal clarification process has been developed specially in the Gordon Lab. The human fecal sample is placed in a growth medium and undergoes filtration for the large particles. The sample is centrifuged with steel beads to break up clumps of cells, and then the sample is suspended in media to preserve cell integrity. As with fecal transplants in humans for the treatment of *Clostridium difficile* infections, the entire fecal sample must be used—it is still unclear which bacteria (and most probably communities) work together.[39] The complex interplay of humans, mice, microbes, and feces is what is under investigation. Human feces (microbes) go into mouse guts; mouse feces (human microbes) come out. While Vanessa Ridaura, Sabrina Wagoner, and Vitas Wagner label tubes, rip pieces of foil, and wipe down surfaces, Philip Ahern muses, "Wouldn't it be more relevant to human biology to study mouse microbiota in *mice* than human microbiota in mice?" This echoes Nelson's question in *Model Behavior,* "what exactly do these researchers think their models are (or are not) useful for, and how do they manage the strength of associations they make between animal and human, behavior and gene?"[40] The mouse outcomes are translated into large sets of genomic data, producing direct correlations between malnourished human microbiotas and malnourished phenotypes in mice. Yet germ-free mice occupy a liminal place between model organisms and biological life that exists only as produced in the lab. In addition to bracketing the age-old question of whether what happens in mouse biology (mouse microbiomes) is generalizable to human biology (human microbiomes), scientists must contend with a further dislocation: the uncertainty of microbes' role in conjunction with environment and diet and how experimental results will map back onto human health.

In a few minutes, the mice will be sacrificed, or "sacked," by cervical dislocation (a rapid separation of the spinal column from the brain), and the four scientific members of the group will become a flurry of dexterous hands. They quickly split the mouse body down the middle and pull off mouse skins in one piece, dissecting the gut into parts that are differently important to different experiments: lymph nodes, spleen, colon, cecum, stomach, pancreas, liver, intestine, fat pads. All the parts are weighed, measured, examined. Vitas has the tiny stomachs and intestines on a cutting board and is pushing feces out with a miniature glass tube. After the intestine is flushed for any remaining cells, Vanessa cuts it into one-centimeter pieces, submerges the pieces in a beaker with phosphate-buffer solution, and drains them again to dislodge epithelial cells. The mouse becomes mouse parts, becomes organs, begins to shrink. There is surprisingly little blood. Someone comments, "You guys have no respect for host biology," and everyone laughs. They recognize that the mice are taken for granted; right now, it is the microbes (in feces, tissue, epithelium) and what the microbes do (in fat, immune cells, and blood) that matter.

The study of human microbiota by scientist or anthropologist requires a continuous shift of scale. In the Gordon Lab, the sheer volume of data being produced by the shotgun-sequencing reads of microbial genomes is unparalleled (and somewhat uncontrollable), requiring the invention of new computing technologies, storage facilities, and scientist skill sets. The genomic DNA sample is first fragmented into a library of small segments that can be uniformly and accurately sequenced in millions of parallel reactions. The newly identified strings of bases, called *reads,* are then reassembled using a known reference genome as a scaffold (resequencing), or in the absence of a reference genome (de novo sequencing). The full set of aligned reads reveals the entire sequence of each chromosome in the genomic DNA sample.[41] At the other end of the spectrum, the objects being studied are the smallest (gene expressions of microbes) and least powerful (fecal samples of infants in resource-poor countries).

After the sack, I spend the rest of the day back in the Gordon Lab at Vanessa's bench with her and Philip, processing the samples. Philip is the lab immunologist, and immunology is what requires the gut to be segmented into so many parts, is what compels the sack day to be one continuous, painstakingly linear process of reducing the mouse body to cells. The small intestine has the greatest diversity of cells, and DNase further breaks down the tissue so that they can look at the cells they want. These protocols are developed by research scientists and then sold to labs in "kits" that include required enzymes and other supplies. But people in the Gordon Lab have adapted the protocols to work better, altered them to be more efficient. Some still have hundreds of steps. Chunks of mouse intestine are stirred for twenty minutes, strained with a kitchen strainer, and come through the other side of a sieve a pinkish liquid. Vanessa rinses everything, to get as many cells as she can from the glass smasher and beaker, and it's strained again through a nylon filter. Philip centrifuges the sample, the cells separate into a sediment, and they purify cells even further with a polymer. Another centrifuge separates epithelial cells, lympocytes, and dead cells into top, middle, and bottom layers, though as with most steps in this protocol, to me it looks like they are just pipetting clear liquid from one tube to another. Michelle Smith, whose bench is nearby, comments, "When you first start doing molecular biology, you just go on faith because you don't see any pellet." They have to trust that cells are collecting. More pipetting, and the sample comes out of the centrifuge again, this time with a cloudy ring in the middle. These are the cells. With a tiny, flexible pipette, they suck out the ring of cells, add more liquid (buffer and nutrients), centrifuge again, and plate in wells. At the end, after the last centrifuge, there are dark particles

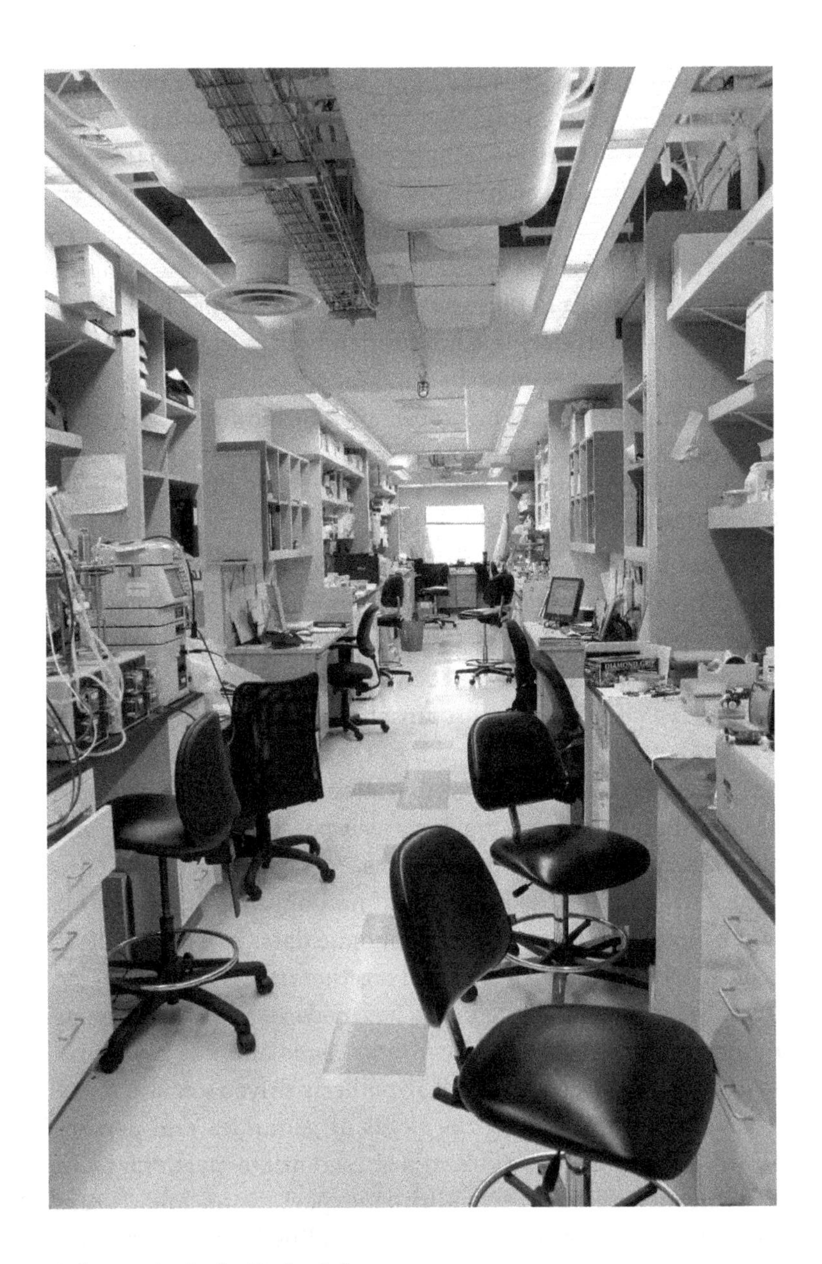

Early morning in the Gordon Lab.

at the bottom of the wells and liquid on top. Vanessa quickly flips the plate upside down into the sink to get rid of the liquid—the DNA of these cells will be shotgun-sequenced to identify the populations of microbes present, and they will also undergo a series of steps in an RNA protocol to see the gene expressions of the microbes and host in each different part of the gut. Twelve samples, twenty steps, and 240 tubes later, the scale of biology is shifting. A phenotype of obesity in a Missourian twin is translated into a mouse body through microbes and food. Mouse becomes anatomy, becomes fleshy chunks, becomes turbid liquid, becomes cells, becomes DNA, becomes RNA, becomes data.

Computational Malnutrition

In recent years, the development of metagenomics and its associated big data technologies have completely transformed how scientists study microbes. Metagenomics allows for vast communities of microbes to be identified through their DNA and addresses a more ecological approach to understanding microbes within their environment and how they function, through a series of experimental and computational approaches centered around shotgun sequencing, massively parallel sequencing, machine learning, and bioinformatics. Metagenomics results in huge data sets, literally hundreds of millions of gene reads. This much biological, genetic, and taxonomic data have never before been analyzed, stored, or accessed together in this way. Datafying microbes through their genomes has made visible a tremendous number of previously unknown organisms and has moved research away from conventional, hypothesis-driven science.

Like many working in the fields of genomics and personalized medicine, the Gordon Lab scientists participate in "promissory science,"[42] building expectations and future contexts for a technology-reliant discipline based on the "anticipation"[43] of a predictably uncertain technoscientific and biomedical future, one that so far exists mostly in speculations and promises. As Hedgecoe points out, this is not to say this practice is particularly unusual or misleading, but the ways in which technological visions are deployed can both direct the paths of research and shape our hopes for this science.[44] As the microbial terrain of the human body is translated into colossal data sets for algorithmic probing, there emerges a governance of health that is increasingly understood through datafication. Current studies of microbiota are attempting to reach beyond DNA to characterize the expressed functions of microbe communities by using "genome-scale platforms."[45] The large amounts of data these platforms produce can be seen to be making what Adams, Murphy, and Clarke call speculative forecasts,[46] defining the present of microbiome science through its anticipated future and creating real, material trajectories where health states and treatments will be diagnosed and prescribed based on the functional genomics of one's personal microbiota.

The development of new technological platforms and the imagining of new research questions emerge as a multidimensional coproduction. Here the plausible interventions that might flow from datafied answers loop back on the mobilization of old and new technologies to innovative ends. Computational biology and bioinformatics enable microbiome research, but the data management and analysis systems for microbial metagenomic studies also influence research trajectories toward the molecular basis of disease. Genomic technologies and high-throughput sequencing, such as those used in metagenomics, have transformed what it means to be a biological scientist. Big data innovations have changed what sort of knowledge is both knowable and valued, coemergent with new modes of postgenomic investigation[47] that have transformed the ways in which science can think about multicellularity, individuality, and organisms. This technology affects life science debates about biosocial environments and human difference.

Accordingly, computing needs in the Gordon Lab have grown enormously in recent years, coincident with the rapid increase in data that they generate from human specimens as well as mouse models. Gordon Lab members jokingly commented that they never expected to be so knee-deep in computational biology and feces. Each new member of the Gordon Lab receives training in fast-evolving bioinformatics and next-gen sequencing methods. They will spend their lab years in a highly interdisciplinary environment, often at considerable distance from their conventional training in biology, genetics, or immunology. The analytic tools used for deciphering the immense data sets from human gut microbiota make implicit assumptions about what the data say about genes, microbes, and human health, and the implications of these developments are subject to internal debate. There are plenty of pertinent concerns about overhyping the role of microbiota in human health. While the metabolic networks of microbiomes may become discernable through next-generation sequencing technologies, too little is known about how microbial genomes and communities interact. Asserting causation by any one bacterial species could be very difficult to prove; more is needed to know what is true and what is false in the data.[48] According to some, the vastness and complexity of microbiome networks and their interactions need to be characterized before causation can be determined. Members of the Gordon Lab were very careful about making any ambitious microbiota or health claims and urged caution against any prescriptive recommendations. Microbiome scientists also have concerns about contributing to the already huge and largely unsubstantiated commercial probiotics market (discussed more later). The new tools and creative experimentalism of microbiome datafication open up discussions of limits and horizons of possibility.

Over the last few years, malnutrition has become a major focus of the Gordon Lab, and scientists are trying to determine what metagenomic studies of gut microbiota say about the causes of wasting, stunting, and undernutrition. Their working hypothesis situates the microbiome as a metabolic organ—the functions the microbes perform, what they do with food on a molecular, biochemical level, and how they interact with human genes—reflecting what Landecker has described as a turn in life science toward a newly revitalized metabolic biochemistry.[49] Yet microbes characterized through functional genomics and metabolomics create critically different bioinformatic considerations between data and lived lives. As Landecker considers in her brief history of food chemistry, *Being and Eating*:

> hunger and obesity are complex and devastating diachronic legacies of intercalating starvation and abundance: the history of nutrition and therefore nutrition science are enfolded in the very bodies that science and policy are attempting to control. . . . In trying to figure out what went wrong, the nice equivalencies of you are what you eat begin to sound off, as it becomes harder to reckon both sides of the equation—"you" and "what you eat" are difficult to define, if you contain both generations and multitudes, and what you eat turns out to not be easily equivalent at all to how hungry or sick you are or are not, and to itself contain worlds of industry and production.[50]

As in this keen analysis, subjects from Mirpur (and their multitudes of microbes) enter the microbiome study with a clinical designation of malnourished or healthy, and they "arrive" in the lab represented by their microbiota in fecal samples. The colonial, economic, and political history of undernutrition as a category and of Bangladesh itself is "enfolded" in these newly enrolled bodies. Malnutrition is presumed socially and medically evident, and microbiota is categorized as malnourished before it enters the experimental pipeline. Lab studies have begun to draw direct correlations between malnourished microbiotas and malnourished phenotypes in mice, when the mice were also fed diets made in the lab to embody local foods. Whether and how malnourished microbiota is transformed by different diets is measured through molecular effects; this take on human hunger proposes to intervene at the level of one's microbiota, read through data.

As mentioned, using human microbiota in mouse guts shows a replication of human phenotypes in mice. For instance, humanized mice fed *Western* (high-fat, high-sugar) diets have increased adiposity;[51] fatness is seen as transmissible via microbiota transplantation. In turn, "personal culture collections" can be cultured from human samples and put in germ-free mice to define the effects of diet manipulations.[52] These effects of diet-by-microbiota interactions can be seen as transmissible and modifiable.[53] Germ-free mice are humanized with "malnourished" microbiota and fed diets based on foods made to approximate their nutritional localities. By analyzing the resulting and presumably locally specific microbial genomics, microbes and their behaviors are turned into computational data—namely, the characterization of bacterial species and their gene content—through which scientists find relationships between food and functional microbiota. As germ-free mouse models reduce microbial relationships to their biologic components, the Gordon Lab attempts to set up a translational medicine pipeline: human–mouse–human. Multiple forms of data are produced from the fecal samples; human microbiota studies in mice are rendered into medical interventions for people, treatments meant for the microbes inside humans. This shows the extent of necessary and complicated cross-species and cross-disciplinary translation in human microbiome studies.

As the molecular effects on/by microbes are studied through computational data, the material, daily complexities in which undernutrition takes shape—food access, sanitation, and poverty—become obfuscated in shotgun-sequenced reads. Medical anthropologist Ian Whitmarsh has pointed out that in postgenomics, the relevance of the genome comes only at the moment of its contact with one of many competing environments.[54] This interaction brings genetics to the forefront of public health concerns but, at the same time, limits "environment" to factors that have a measurable effect on genes. What the data from mice guts reveal about microbes allows scientists to make conclusions about how diet, human/microbial genetics, and environment work together. Mice are the manipulable flesh of the experiments, vehicles for data and potentially transformative agents. But like Whitmarsh's Barbadian asthma, the experience of Bangladeshi malnutrition is translated into a microbial genomics of locale, ethnicity, and culture. Gene–environment interactions become a tool for researchers to account for disparities in microbiomes: the dependent, plural relationality of "nature" (genes) and "culture" (environment). Lab thinking participates in the kind of work described in Landecker's analysis of nutritional epigenetics,[55] where food is understood as bioactive molecules that become part of the molecularized environment in which metabolic systems operate—and social difference embodied in genes is open to environmental intervention. For Landecker, race, gender, and class are the genetically embodied socialities influenced by environment. In the Gordon Lab, it is Bangladeshi urban poverty. Furthermore, in her studies of the increasing focus on gene–environment interaction in environmental health sciences, sociologist Sara Shostak has described how environmental exposures and biological and social "somatic vulnerabilities" create categories of risk inside and out, through the inextricable interaction of genes and world.[56] Mothers and children in Mirpur are somatically vulnerable to enteric disease, and datafication determines the microbial risk of undernutrition on a molecular level.

In recent years, data-driven enactments of malnutrition in the Gordon Lab have focused on the concept of microbiota "maturity."[57] By applying a machine learning–based approach to 16S rRNA data sets generated from fecal samples obtained from the Mirpur children, scientists identified bacterial configurations whose proportional representations define a healthy gut microbiota as it assembles during the first two years of life. In other words, they defined "healthy maturation" of the microbiota through specific bacterial species in children who exhibited consistently healthy growth (by WHO reference standards). Healthy kids equal healthy microbiomes. Then, these age-discriminatory bacterial species were incorporated into a model that computes a relative microbiota maturity index and microbiota-for-age Z-score. By representing the different stages of assembly of a "healthy" Bangladeshi child's gut community, they could compare the microbiota of children who exhibited physical signs of malnutrition. What does the microbiota of malnourished kids look like compared with a healthy microbiota? They found that severe undernutrition in children is associated with what they called *significant relative microbiota immaturity*; that is, the malnourished Bangladeshi children had a microbiota configuration more similar to younger Bangladeshi children than to healthy kids their own age. There is an oscillation here between datafied microbes and conventional, clinical diagnoses of malnutrition. Health status is predetermined by observed height, weight, and growth—malnutrition is not decided by microbiota, but microbes can be algorithmically and accordingly designated as immature, hence malnourished. These are the categories of malnutrition and health that emerge as big data becomes a tool for nutrition science.[58]

Historically, malnutrition has been understood in terms of physical and cognitive growth retardation, but this measurement provides a new view of human postnatal development, one in which the child's microbial populations are those that are stunted. Therapeutic food interventions only "fixed" the microbiota while the foods were being eaten; when treatment stopped, the microbial populations shifted back to a more immature state. Identifying this process has led the Gordon Lab to propose that persistent microbiota immaturity is not compatible with healthy growth, and the focal point of their work has become finding new means for more durable rescue from what scientists call microbial "developmental delay." Entangling scientific discovery, parental responsibility, and child development, this language is a timely microbial echo of neurodevelopmental concerns in an age of autism.[59] "Rescue" of the microbiota and, consequently, of the child could come in the form of food-based interventions, conditioning of the microbiota for a more favorable response to diet, or personalized probiotics. A Gordon Lab paper by Subramanian et al. suggests applying a developmental biology model to ask questions about persistently immature microbiota:

> One question is whether the developmental program defined in Bangladeshi infants and children is generalizable to other populations representing different geographic and cultural settings. If so, it would reveal a fundamental shared aspect of postnatal human development and raise mechanistic questions about the factors that specify a healthy microbial community "fate."[60]

The discovery and pursuit of the idea of microbiota maturity place emphasis on the (generalizable) molecular, microbiotic body and, simultaneously, on the translational pipeline. Microbial fates determine human health outcomes; our microbes are as unique as fingerprints yet universally functional. The direction of the Gordon Lab's work has undoubtedly been influenced by the support of one of the world's largest private health foundations and the joint mandate to address malnutrition as a global problem. There is a sense of responsibility to consider fast-tracking useful results in the form of translational applications. Such findings rely on identifying human states of health through microbial biomarkers that are culled from next-generation sequencing technology. The research also seeks implications in the technology that will become interventions—be it through prolonged administration of therapeutic food, the prescription of ingestible microbes, or what the lab calls "next-generation probiotics" using gut-derived taxa aimed at repairing the community by assisting microbiome maturation.

Malnourished states are enacted through bacterial taxonomic biomarkers. Thus data are operationalized through local microbiology, seeking standardized health interventions—"systematic analysis of microbiota maturation in different healthy and malnourished populations living in different locales, representing different lifestyles and cultural traditions may yield a taxonomy-based model that is generally applicable to many countries"—or a more personalized one, based on diet and environment: "alternatively, these analyses may demonstrate a need for geographic specificity when constructing models (and diagnostic tests or therapeutic regimens)."[61] The influence of funders driven by the goal of rapid translational medicine and focused on a regionalized, easily deliverable malnutrition solution is seen in such imagined uses for new technologies.

Datafuturity

For the Gordon Lab, interpreting new big data is a means to seeing people as human–microbe composites and is based on a supposition that the actions of our microbial partners are accurately deciphered through metagenomics. The dream of microbiota treatment would be personalized diet interventions based on what lab scientists call *personalized gnotobiotics,* germ-free mouse/piglet models colonized with an individual's microbes from which the interrelationships between the host's physiology, microbes, and food could be explored. *Personalized nutrition* would be based on diet-induced microbiota alterations that cause changes in host physiology, including disease development and progression.[62] Between 2009 and 2021, investigators in the Gordon Lab moved purposefully toward the goal of customized medicine based on metabolomics and genetic testing, seeking to manipulate "unhealthy" or "immature" microbiomes by prescribing next-generation probiotics and synbiotics (mixtures of probiotics and prebiotics that beneficially affect the host by improving the growth or activating the metabolism of "health-promoting" bacteria) to be ingested as food supplements. Yet such precision microbiome manipulation is still only aspirational in terms of new technologies and new actions: "major limitations in 'big-data' processing and analysis still limit our interpretive and translational capabilities concerning these person-specific host, microbiome and diet interactions."[63] These methods are still subject to all the instability to which such dreams have succumbed in deriving many straightforward health interventions from mapping human genomes.

Precision medicine begets a specific type of gene-based, individualized intervention. Microbiome science has followed suit, endeavoring to create personalized microbiomics, focused on addressing health from a microbial perspective—an eventual microbe-based therapeutic intervention that would customize medicine/probiotics/food/health care based on proteomics, metabolomics, and genetic testing. These "inside-out" approaches are taking the interventional form of *microbiota-directed foods* (MDFs) or *microbiota-directed complementary foods* (MDCFs) and *next-generation probiotics* (human gut–derived microbial strains) designed to repair microbiomic abnormalities or deficiencies.

After showing that human microbiomes could be studied and manipulated in germ-free animals and that a malnourished microbiota was an immature microbiota, the Gordon Lab took several steps in quick succession toward a translational medicine microbial intervention. All these studies used Mirpur babies and their microbes. In 2019, the lab published work in *Science*, "Effects of Microbiota-Directed Foods in Gnotobiotic Animals and Undernourished Children,"[64] which showed research that humanized gnotobiotic mice and piglets with malnourished Bangladeshi microbiota and then looked at the host and microbe changes with MDCFs. Their conclusions were not just that the immature microbiome could be altered with MDCFs but, boldly, that a healthy microbiota is essential to healthy growth and development. This paper was followed closely by a published study protocol, "Proof-of-Concept Study of the Efficacy of a Microbiota-Directed Complementary Food Formulation (MDCF) for Treating Moderate Acute Malnutrition."[65] This showed preliminary research and the design for a prospective study to determine whether MDCFs would improve weight gain, microbiota repair, and plasma biomarkers/mediators of healthy growth over RUTFs. There's an incremental shift in language between the two papers, where "what will MDCFs do?" becomes "MDCFs are a treatment for malnutrition." This study resulted in "A Microbiota-Directed Food Intervention for Undernourished Children,"[66] published in the *New England Journal of Medicine* in April 2021. This paper made definitive claims that the Gordon Lab–created MDCF formulations from experiments with gnotobiotic animals "aged" immature microbiota and improved health status biomarkers. Then, in September 2021, all this work culminated in a comment written for *Nature Medicine* by Dr. Gordon, Dr. Ahmed, and other PIs from the Gordon Lab and icddr,b, "Melding Microbiome and Nutritional Science with Early Child Development."[67] This comment focuses the experimental research with MDCFs as a tool to reframe nutrition, "A New Kind of Food Pyramid That Considers the Gut Microbiome as Part of Nutrition Itself."[68] The authors propose that ideas of wellness must now include the developmental state of microbial communities, that point-of-care diagnostics and interventions are dependent on states of both host and microbe health.

New microbiome therapies as health-promoting foods will be designed from the "inside- out" and will take into account the consumer's particular microbiota (profiled through data-driven metagenomics) and their individual ability to transform ingredients into metabolic products. This research direction raises questions of profit and unequal access to and distribution of technologies, as well as the vulnerability of publics to "molecularized marketing."[69] These implications require vigilant monitoring by life and social scientists, especially those working in funded alliances with food companies. Identifying gut microbial populations through datafication (be they "malnourished," "immature," or potentially designated as "obese" or "inflammatory") may lead to emerging medical probiotic markets in conjunction with and beyond global health interventions, bringing together sometimes competing ideas of humanitarianism and economy. Interventional microbiota products may fall into regulatory categories of "supplements," "functional foods," "medical foods," or even "drugs," blurring the boundaries between food and medicine.[70]

Medical anthropologists, ethicists, public health scholars, and Gordon Lab scientists themselves have written about the growing ethical, legal, and social challenges resulting from studies of human microbiota. Dietary supplements resulting from human microbiome research open the door to what they call *commercialized intervention,* "the proliferation of commercial products that claim maintenance or restoration of good health, and prevention of disease or sickness with the use of good bacteria."[71] Implementing Waldby's term, *biovalue,* Slashinski et al. argue that human microbial ecology makes claims about the "biological vitality" of human microbiomes in ways that instrumentalize knowledge about microbes to create commodities to be sold under the guise of contributing to human health.[72] The data-driven conclusions about which microbiota are functionally lacking, or what species may be overabundant and correlative to human illness, propel interventions in food/supplement/diet directions. The Gordon Lab believes that an "evidence-based framework" for classifying these products will help solve this problem; their branding, testing, and labeling all depend on understanding what they actually do. Beyond that, these scientists think that our very definitions of nutrition and health will be transformed, because "microbiota generates biomolecules that are not produced by any of our human cell lineages and that affect our health status . . . forcing us to evolve our concept of essential nutrients. . . . The development of MDFs will likely help change concepts and definitions of nutritional requirements, nutritional benefits and food safety."[73] Technological innovation in datafying microbes makes new microbiome markets possible.

Another good example is the company Evivo—a probiotic-producing corporation cum scientific research institute, part of Evolve Biosystems, founded by human microbial ecology scientists working out of the Foods for Health Institute at the University of California, Davis. Evivo's tagline is "Breast milk is perfect. Evivo makes it better," and the company sells $75 starter kits of infant probiotics (as well as refills—a six-month supply is $335): a supplement of powdered *Bifido longum infantis*[74] bacteria to be added to breast milk and administered to babies continuously as long as they are breastfed. It has sponsored many posts on "mommy blogs" and capitalized on the scientific reputations of its founders. Evivo uses language like "Most moms today don't have the right bacteria to pass on to their babies. Evivo is activated *B. infantis* and works with breast milk to restore the infant gut microbiome to its original, natural state,"[75] calling dramatic attention to the maternal lack in microbial care and the ability of an expensive product to restore health to the infant microbiome. This company has even constructed a new disorder, "newborn gut deficiency," to describe an overwhelming shortage of good bacteria in the infant gut—fixable, of course, by using Evivo's product.

As the biovalue of human microbiome research is in danger of being increasingly assimilated into capital value, public health interests, especially in the case of vulnerable populations like those in Bangladesh, need to be carefully balanced with the scientific research, funder goals, and the industry marketplace. Evoking Sanabria and Yates-Doerr's alimentary concerns, human microbial ecologists are sharply attuned to the care that must be taken:

> These research platforms offer the promise of yielding next-generation foods designed to be satiating, delicious, nutritious, and able to manipulate microbiota and host properties in ways that promote healthy growth and wellness. However, fulfilling this promise demands a holistic view of the nexus of human gut microbial

> ecology research, agricultural practices, food production, evolving consumer tastes in an era of rapid globalization, envisioned commercialization strategies, current regulatory structures/practices, ethical issues, and public education.[76]

For Bangladeshis like Rimi and Kanta, the effects of poor infrastructure (sewage, sanitation, clean water) on human bodies (diarrheal disease and, consequently, undernutrition) may or may not be altered by the scientific focus on datafied microbial bodies. The promises of microbiome research and its imagined solutions to malnutrition venture toward the individualized interventions of personalized probiotics or therapeutic foods, yet in Dhaka, treatments are more likely to be based on geography, genetics, and a collectively shared malnourished or "immature" microbiota. Applied microbiome research occupies a seemingly contradictory ground between personalized genomic medicine and global population health. Just as Rimi shares her two-gas-burner stove with five other women, she shares the position of study enrollee with most of her neighbors. She anticipates the poor health diagnoses of her future children and asks me, if she gets pregnant again soon, can her next baby also be enrolled in the study? Rimi's question speaks to strategies for seeking health care among mothers in Mirpur, who are utilizing the resources at hand to care for their families. Ethnography's "empirical lantern" may be able to shed light on the transnational and local realities of people embedded in global scientific research and health programs[77] with the goal of a more gradated context through which to enact ethical and effective science.

Incommensurabilities engendered between everyday life and data-driven pathways to health intervention create areas of inquiry for scholars using big data to study nutrition or the social determinants of global health, especially in terms of the speed and scale differences between places where knowledge is being made: such as Mirpur and the metagenomic databases of the Gordon Lab. As microbiome science anticipates new ways of seeing malnutrition through microbes that originate in human guts and become datafied through 'omic processes, there are opportunities to begin to explore what new material trajectories, environments, and futures are emerging for researchers and research subjects. The struggle to apply scientific microbiome work to real-world conditions underlines the flux between the "large" social issues and the "small" molecular microbiome, especially through immensely datafied methods and findings.

In 1888, Rudyard Kipling published a short story called "The Phantom Rickshaw." It seems fitting to begin a chapter about the phantom of race with an image that recalls an author who was an imperialist racist colonialist.

FIVE

Ghosting Race

A seemingly never-ending paradox erupts around race in every new area of scientific research: race in science is commonly affirmed as a social construct, but not quite. Sociologist Ann Morning shows ethnographically that despite that common belief that constructivism ruled, much of the scientific literature of the early 2000s was premised on race, "grounded in and reflective of human biology."[1] While some scientific research claims that there is more to human diversity than race, other scientific work finds biological distinctness in race. Categories of race are used unceasingly to arrange scientific subjects and their bodily differences. Sometimes scientists invoke the "social" to avoid the unquantifiability of race in their work, while simultaneously operationalizing race to identify differences in biologies. Often, race becomes a determinant biological category without any interpretation of commensurate sociomaterial factors; race in science fluctuates between solely social and simply biological. Between the 1970s and 2000s, there *seemed* to be broad cross-disciplinary agreement about the social construction of race. After the genomic revolution of the late 1990s made claims about the genetic diversity/relative commonality of genes in humans, a clear debate around race and science ensued—not only between scientists but also among anthropologists, sociologists, and bioethicists. Many scholars have grappled with the critical, empirical, and everyday implications of this fraught biomedical–race relationship.[2] Here I mine this analytical history to create a framework for understanding race at work in research on the human microbiome. There is no agreement (in any discipline) about what race is or where it resides, about what is considered "biological" or "social" or if those things are discrete. The introduction to *Anthropology of Race*[3] describes two different anthropological interpretations of how genetic science approaches race. One perspective presupposes the social constructionist view and forces a focus on racism rather than race, leaving the biological out entirely.[4] The other perspective sees new genomic science as a revival of historical race science.[5] Neither view proposes that it is possible to eliminate race from scientific research. Race becomes a ghost in the scientific work, an invisible, powerful informant that affects the categorization of bodies, how difference is scientifically made and verified, and, ultimately, how interventions and care are applied. But what happens as translational research is increasingly datafied and personalized? Does the ghost of race haunt these new houses?

Biomedical studies of human microbiota and the microbiome gesture toward the postracial aspirations of personalized medicine—characterizing states of human health and illness microbially, dependent on individual biosocial factors like nutrition, diet, and environment. PubMed shows 55,266 human microbiome papers published between 2009 and 2019. I use the findings of the NIH-funded HMP as the baseline of accepted microbiome knowledge. While being mindful of reductive generalizations, here I use a series of tenets in human microbial ecology that are undisputed in the field. By viewing humans as assemblages of different species that form ecological units, some microbiome science attempts to complexify the binding historical categories of race. If humans are made mostly of microbes, which presumably don't have races, then humans' interspeciality or "becoming-with"[6] these organisms takes precedence over old racial categories.

But, inevitably, unexamined categories of race and ethnicity surface in a myriad of studies on microbiota from asthma and diabetes to colorectal cancer and bacterial vaginosis. The bodies chosen for microbiome research are explicitly raced and sexed as a central part of the scientific work. This chapter approaches race as a ghost variable across different kinds of microbiome research. By "ghost variable," I mean politically, socially, and historically charged racial categories that are used in studies of the microbiome without "race" ever being explicitly named. The formulation of race as a ghost variable came out of a Society for the Social Studies of Science conference panel and subsequent *Science, Technology, and Human Values* special issue in which I participated, organized and edited by Katrina Karkazis and Beck-Jordan Young:

> A chief appeal of the metaphor of the ghost is that it brings the importance of history to the fore. Ghosts are simultaneously history and the present, not just an accretion of earlier experiences, but the palimpsest left when one tries to erase them. Sometimes faint and hard to discern, sometimes rambunctious and disruptive, ghosts refuse our attempts to simply move on.[7]

Slightly different from Subramaniam's ghosts in science, which she is attempting to consciously coexist with, these ghosts of race are deliberately obscured. So, this chapter asks, What is race doing in studies of the microbiome? Why is it there, and how is it functioning? Scholar of race, gender, and law Dorothy Roberts suggests that the way out of this race–science paradox is to focus on how race is being used as a categorizer, instead of trying to find differences in the biologies of race.[8] I examine microbiome research to argue that social scientists and STS scholars must work with biological scientists to put microbial differences into perspective—to investigate how these differences are informed by biosocial determinants more complex than race alone, and how race is never alone, never an independent explanatory variable. In 2013, anthropologists at the School for Advanced Research attempted to reorient problematic anthropological perspectives on race and posited that race is recursive: "the assertion that race is a biosocial fact, indeed, moves us from thinking about it as a fixed quality, an inherent essence, or a unique causal mechanism to seeing race as a process, one that offers no certain line by which either 'biology' or 'society' can be starkly delineated."[9] Following the work of Mukhopadhyay and Moses (unified biocultural approach)[10] and Jackson (ethnogenetic layering),[11] they strongly emphasize turning anthropological attention to sociocultural and biological data on race. An anthropology of microbes regards microbiomes as entangled embodiments of what are considered social, environmental, material, and biological elements, but requiring us to reckon with human–microbe relationships as intra-acting biosocial assemblages.[12] This chapter advocates for the continual unsettling of the social–biological divide in terms of the microbiome and race. Ultimately, transdisciplinary collaboration is required to address racial health disparities in microbiome research without reifying race as a simple designation, and to study the biosocial intersectionality of the human microbiome.

As microbiome research becomes one of the most popularly publicized areas of scientific work, race functions as a ghost variable in categorizing human bodies and their corresponding microbial partners. "Race" as an operational concept in microbiome science has a ghostly presence, one that is there but not there, hiding in shadows and jumping out when least expected. M'charek, Schramm, and Skinner have described this ephemerality as race's absent presence,[13] where race is a slippery object. These authors go on to say that the contemporary liberal politics of racism (and I would include neoliberal science) often *means* race but codes it as national, cultural, or religious identity. Following this, many studies of the microbiome vacillate between the terms *ethnicity, geography,* and *genetic ancestry,* but all serve as a cipher for historically and culturally saturated designations of race. In this chapter, I look at race/ethnicity-dependent areas of microbiota research: in the microbes of "uncontacted hunter-gatherers,"[14] vaginal microbes across ethnicity,[15] and my own ethnographic fieldwork in the Gordon Lab in studies comparing guts across the globe.[16]

I use this research to think about what is elided when human microbial ecologists make biological microbiome claims presuming racial categories and how the inclusion of social science thinking can perspectivize microbial differences within poverty, resource access, and discrimination. In the *mSystems* article "Chasing Ghosts: Race, Racism, and the Future of Microbiome Research," my coauthors and I continue to evolve the ghost variable concept. We outline ghost variables as racialized terms that serve as oversimplified proxies for racial and/or racist structural drivers and their associative factors.[17] In a review of the microbiome literature since 2000, we found many behavioral, geographical, political, and environmental ghost variables (underdeveloped, inner-city, rural or urban, genetic ancestry, etc.) that obfuscate associative factors, which are usually categorized as social determinants of health. Furthermore, we argue that these associative factors (access to clean water and sanitation, food security, climate differences, toxic exposures and pollution, access to health care, immigration or war trauma, etc.) are products of structural drivers. These structural drivers include capitalism, industrialization, environmental injustice, systemic racism, heteropatriarchy, imperialism, settler colonialism, and resource exploitation. Because structural drivers are rarely addressed as causal factors (or at all) in scientific research, social science interventions are needed. Ethnography can be a critical first- and second-order analytical approach in investigations of how social determinants of microbiomes are also biological. Anthropologists can provide scientists with qualitative ethnographic data about living conditions, social networks, daily practices, and "local biologies,"[18] exposing how biological and social lives are mutually constitutive over time in what Lock and Nguyen call biosocial differentiation.[19] This chapter serves as an initiatory interrogation of race in microbiome science not only to see how race and racism function within current research but also to think about who microbiome science is for and how to counteract the structural violence built in to its technologies and aims.

What Race Means

Because race is formally scientifically designated in various ways, it is important to start with categorical clarity. How funding and regulatory agencies like the NIH, the NSF, and the U.S. Census

Bureau define race and ethnicity becomes the ways grant proposals, scientific research, and scientists use and consequently produce race and ethnicity. As M'charek discusses, the conclusions drawn about genetic (racial) difference and the technologies created and used to define those differences are epistemologically entangled.[20] The NIH race and ethnicity standards are set by the U.S. Office of Management and Budget and take an ancestry/country of origin approach ("a person having origins in any of the original peoples of . . ."), where races are American Indian or Alaska Native, Asian, Black or African American, Native Hawaiian or Pacific Islander, white, and Hispanic or Latino (ethnicity regardless of race).[21] Unfortunately, what is meant by "origins" and "original peoples of" is not well defined. The NSF uses the same categories but allows for respondents to choose more than one designation. The NSF also explains the many problems with reporting race: populations change over time, self-identification is inexact, and heterogeneity in populations can be overlooked. Because nearly all the microbiome studies discussed here (with the exception of the Gordon Lab work) are federally funded by the NIH or NSF, and all the research takes place in U.S. laboratories, these are the parameters for race and ethnicity that I will use throughout. Microbiome studies often conflate, confuse, and interchange ethnicity, nationality, and geography with race.

While social scientists work on constructing new research language,[22] federally funded science looks to technology to circumvent the race problem. Precision medicine's "post-racial promise"[23] has the tenuous potential to make health care more data driven and biologically accurate, though those objectives do not necessarily prioritize or track with racial equity.[24] Precision medicine begets a specific type of gene-based, individualized intervention. Scientists working on the microbiome have followed suit, endeavoring to create personalized microbiomics, addressing health from a microbial perspective. As discussed in chapter 4, goals include microbe-based therapeutic interventions that would customize medicine, probiotics, food, and health care based on proteomics, metabolomics, and genetic testing.

Looking at states of human health and illness through a microbiomic lens disarranges conventional categories of community, species, and self. Social scientists must be attuned to when the ghost variable of race begins to materialize and read microbiome studies across the grain of racial categorization and inequity. Stefan Helmreich is one of few anthropologists beginning to speculate about race in the microbiome. In *Sounding the Limits of Life,* he comments that even though the microbiome seems to be able to create categorical havoc, ultimately, "there is nothing preventing race from manifesting in microbiome talk, in both reductionist and complex ways."[25] Although I share Helmreich's caution, and show later how race does surface in microbiome work, I propose that cross-disciplinary cooperation can disrupt this seeming inevitability. I analyze three specific research areas in microbiomics and take Helmreich's challenge as an opening instead of an end point, turning his statements into a series of questions: How does race manifest microbiomically, and to what end? What social science interventions can help to ghost-bust these manifestations of race?

Uncontacted Microbiota

Technically, microbiome science is race-free, but subjects of microbiome research are often placed in familiar, opposing groups: "Westerners,"[26] who are primarily white and are assumed to have similar lifestyles and socioeconomic statuses, versus Black and brown bodies in the Global South, assumed to be underdeveloped or "modernizing."[27] The American Gut project, based out of the Knight Lab at the University of California, San Diego School of Medicine, is a citizen-science, crowdsourced attempt to create a massive public microbiome data set. The first 2018 results showed that participants have been overwhelmingly (87 percent) white, 47 percent with a graduate or professional degree, and mostly in the above $100,000 per year income range—speaking volumes about who represents the "Western" gut. Microbiome differences are sought between Western and developing, without corresponding investigations into existing economic, political, and health vulnerabilities. Race surfaces here, as those historically biomedically exploited become ready bodies for microbiomic explorations.

In June 2019, the Sonnenburg Lab at Stanford University published an opinion piece in *Nature Microbiology* titled "The Ancestral and Industrialized Gut Microbiota and Implications for Human Health."[28] This essay takes a catastrophic view on the damage done to microbiomes by modern living, which the authors refer to as "industrialized microbiota"—microbiomes altered by antibiotic use, increased sanitation, cesarean sections, and industrialized food production. Industrialized microbiota is contrasted against microbiota collected from "traditional populations," which are reported to contain a high abundance of rare microbes. The Sonnenburgs define these microbes as VANISH taxa: volatile and/or associated negatively with industrialized societies of humans. They warn that the loss of these microbes is causing widespread chronic disease and dysbiosis: microbial imbalance or impairment. They propose that "healthier" diets, probiotically engineered foods, and "rewilding" (the practice of reintroducing "lost" bacterial species to the gut) can help save the industrialized gut.

So, what do these prescriptions have to do with race? These authors define "industrialized," but never say what they mean by "traditional populations," so the phrase becomes racially coded to mean Indigenous, undeveloped, and not white. At times the Sonnenburgs are talking about lifestyle and diet differences (foraging or rural agricultural practices, sanitation, unmedicalized births) but call those differences "nationality," "geography," and "race." Here language matters—rewilding calls up racist stereotypes of Black and Native American people as wild, savage, and uncivilized. Rewilding also speaks to a nostalgic return to a mythologized "wild" nature.[29] Wildness is transposed from human bodies onto their microbes and back again. This is an important opening for social scientists to try to understand (and help scientists understand) who the participants are and what categories of race do in microbiome studies. So much is connotated in labels of "traditional" and "modern," whole world colonial histories and global economies. Legacies of nationhood have even made their way into the names of microbes themselves—in one 2020 study, scientists named subspecies of the extremely common human gut bacteria *Eubacterium rectale*: ErEurasia, ErEurope, ErAsia, and ErAfrica (e.g., *E. rectale* Eurasia).[30] Bafflingly, the study insists on maintaining continental divisions for these microbes, describing them as "geographically stratified subspecies," but then goes on to state that in fact, ErEurasia and ErAfrica were commonly found in samples outside their ascribed geographic areas. The authors admit that the functional and genetic characteristics of the subspecies are probably determined by host lifestyle but that "there are no large datasets that contrast individuals from the same population living different lifestyles."[31] In the absence of data, easy global categories prevail.

Salvage ethnography refers to the recording of the languages, rituals, and so on of "disappearing" cultures—those presumed to be under threat of extinction from modernization. Modern anthropology has eschewed this practice, and many anthropologists leveled their criticism at the same kind of work being done in the Human Genome Diversity and International Hap-Map projects. These projects focused on Indigenous populations as valuable to furthering genome science but fell short on equitable practices of consent, community participation, and benefit.[32] In the 1990s, salvage ethnography became a sort of salvage genomics and is becoming what I would call salvage microbiomics, with a repeating loop of the same populations as scientific subject/objects. Salvage microbiomics wants to save valuable, vanishing microbes from modernization without acknowledging microbiome science's own embeddedness (and complicity) in technoscientific systems responsible for changes in people's microbial populations.

In the Sonnenburgs' essay and many other studies, the racially/ethnically defined "traditional" microbiomes are compared directly against "Western" ones to establish stark contrasts—racial othering reflected in microbial difference. These papers appear in high-impact scientific journals like *Nature*, PNAS, and *Current Biology* and become foundational for further studies. Without careful unpacking and recognition of colonizing scientific histories, microbiome science racializes and discriminates. Conceiving of Indigenous people as primordial, barbaric, and undeveloped has long been a driving justification for colonization, enslavement, and genocide. Anthropologist Cori Hayden has described bioprospecting in terms of the extraction of local plants and Indigenous knowledge;[33] here microbes from Indigenous guts by U.S. research labs and corporate partners are bioprospected. "We" have damaged our microbiomes through the overuse and abuse of medical and nutritional technologies, and our salvation will be to return to preindustrialized microbiota. Seeking answers to current Western woes in the idealized purity of the past and primitive gut in turn instrumentalizes brown and Black bodies in the service of white health. Interventions to save the Western gut function differently than those focused on addressing the health crises of American Indigenous populations or people in the majority world, as discussed later.

In the *Science Advances*[34] 2015 Clemente et al. study "The Microbiome of Uncontacted Amerindians," researchers characterize the fecal, oral, and skin bacterial microbiome of members of a Yanomami Amerindian village in Venezuela and describe their subjects has having "no documented previous contact with Western people."[35] Although the Yanomami in question lived in an isolated and unmapped (to foreigners) area, "first contact" is a myth that has long occupied the imaginations of social and biological scientists seeking "uncontaminated" study populations. Anthropology has its own ethically violent history with the Yanomami, and microbiome science has returned to these people seeking microbiomic purity in the same ways that anthropologists sought cultural purity. The fact that the Yanomami sampled were seen as an untapped treasure trove of microbiota further reinforces the idea of salvage microbiomics. Describing Yanomami microbiota as occupying a different temporal space than modern Western microbiota elicits the same othering imaginaries in which majority world populations represent the West's lost past. The idea that "Yanomami microbiota" or "modern Western microbiota" is definable and homogenous is also something that deserves interrogation. In "Reviving Colonial Science in Ancestral

Microbiome Research," cultural studies scholar Stephanie Maroney keenly analyzes the problem:

> These studies assume that peoples living so-called "hunter-gatherer" life-styles in places like eastern Tanzania and the Venezuelan Amazon are appropriate biological proxies for humans living 10,000 years ago. This temporal collapse reduces Indigenous and rural people living traditional lifestyles to mere research fodder—or "living fossils," it depoliticizes their existence in the present by writing of them as untouched and uncontacted.[36]

Included in the ELSI (ethical, legal, and social implications) section of the NIH HMP, specific attention was paid to the ethical challenges resulting from the inclusion of Indigenous communities and the risks to historically vulnerable populations. The main issue is the lack of clinical applications that will benefit these populations in the foreseeable future. What's at stake and what's to be gained by the Yanomami participants for contributing their microbiota? The call for attention to these types of research inequities isn't new—many have speculated about how clinical research should be both ethical and responsive to the specific needs of those in resource-poor countries.[37] An extensive literature on benefit sharing in global health research has emerged, but the conversation has taken place primarily among anthropologists, bioethicists, and public health scholars, not research scientists. Since direct, immediate intervention is unlikely, some propose an "ethics of care" that requires microbiome researchers to attend to the current predicaments of research participants, support meaningful infrastructural changes, and remain alert to possible commercial exploitation. In 2016, researchers at the University of Oklahoma published "Gut Microbiome Diversity among Cheyenne and Arapaho Individuals from Western Oklahoma"[38] in *Current Biology,* based on a study in which the scientists established an interdisciplinary partnership between the university and Cheyenne and Arapaho tribes. They attempted to diversify population-based microbiome knowledge (which had been limited to European Americans) to address gut microbiome–associated complex diseases that are also common health disparities among American Indians. The study focused on how the biological interacts with the environmental and socioeconomic. The primary investigators on the study were both molecular and sociocultural anthropologists, and they employed what they called an “embedded ELSI approach,” one in which the scientific community engages in long-term relationships of mutual benefit and concern, trust, and understanding with the participants from American Indian communities. This study, though not perfect (the analysis of metagenomic data still overshadows the ethical innovation), is a good example of the social and biological science partnerships necessary for successful microbiome science.

Vaginal Sites

The influential 2011 *PNAS* paper “Vaginal Microbiome of Reproductive-Age Women”[39] established a categorical precedent in vaginal microbiome research, claiming definitively that Black and Hispanic women have different vaginal microbes than white and Asian women. The study concludes that these differences in microbes have led to different disease rates; Black and Hispanic women have more bacterial diversity in their vaginas, which makes them more susceptible to bacterial vaginosis (BV) infections. This paper is the most often cited paper on vaginal microbiota, and it laid the groundwork for every subsequent study in this area. Jacques Ravel’s research was funded by the NIH, yet the study uses “ethnicity” as an umbrella term to describe three NIH-defined races and one ethnic group (Black, white, Asian, and Hispanic). Race and ethnicity here are conflated and overlap, underlining colonial histories of difference. In Ravel et al.’s study, ghostly race hides within racialized ethnicity. Seeking a relationship between ethnic background and vaginal bacterial community composition is foundational to the study—race serves as an organizing research principle. With the insights made possible by ever more sophisticated biological and statistical theory, next-generation bacterial genome sequencing, and formidable computing power, we seem still trapped in Linnaeus’s original race scheme, dividing the world's populations into a four-part color wheel.

The Ravel et al. paper redefines what sorts of bacterial communities are found in "healthy" women so that risk and diagnosis can be assessed individually. However, associating Black and Hispanic vaginal microbiomes with BV, which is itself associated with risk factors of multiple sexual partners, lack of condom use, and smoking, leans on racist stereotypes of hypersexualized Black and Hispanic women. Setting up a different "normal" for Hispanic and Black women's vaginal microbiota is a dubious use of questionably related racial identifications. As geographers Mansfield and Guthman have written, new ways of thinking of biology as plastic changes concepts of difference, normality, and abnormality.[40] Microbiomics, just like epigenetics, blurs distinctions of inside–outside and biological–social in nondeterministic ways:

> So while it might seem that these new epigenetic models of plastic life should eliminate race by eliminating notions of discrete kinds given in nature, it appears that epigenetics offers a new form of racialization based on processes of becoming rather than on pre-given nature.[41]

Similarly, because microbiota is open to intervention and optimization, making racial distinctions in microbiomes reinforces normalcy tied to race. But race itself doesn't do much work to help us understand variations in microbiota, making it harder to navigate health disparities without reifying race.

Ravel et al. conclude that the reasons different ethnic groups have these vaginal microbiota differences are unknown, but "it is tempting to speculate that the species composition of vaginal communities could be governed by genetically determined differences between hosts."[42] In a personal communication, a senior microbiome scientist told me that Ravel's connection between microbial communities and ethnicity was unsubstantiated and that when the study was replicated at a U.S. university among undergraduate women across race and ethnicity, it showed the subjects had mostly the same microbiota. This scientist concluded that class and age were more influential factors than race in determining microbial populations in the vagina. In the Ravel et al. research, complex socioenvironmental components are only briefly alluded to and considered separate from the biological microbiome. For example, specific sex practices have an enormous influence on microbial populations in the vagina. Yet monitoring a study subject's sexual behavior and determining what acts at what time on what bodies affect the microbiome is impossible: several sex acts can be concurrent; time of sampling is challenging to manage; and other variables, such as lubrication and hygiene, confound results. So, sex practices that are instrumental in making the vaginal microbiome are hard to study. On the other hand, "race" itself *does not* have an effect on vaginal microbes, but unlike investigating sex practices (difficult), race is easy to assign. BV itself is notoriously ill defined. It would be more accurate to say that BV is more prevalent in populations of low socioeconomic status, specific reproductive age, inadequate nutrition, and so on rather than in Black and Hispanic women. Indeed, there is an enormous public health literature substantiating that Black women are at higher risk of bad maternal and reproductive health outcomes. How can ethical and effective interventions help solve these health inequities without returning to racist categories?

Although some microbial ecologists working on the vaginal microbiome focus more on how age and place within reproductive life are more influential in determining microbial communities,[43] the novelty and assumed greater precision of metagenomics have eclipsed an examination of the socioenvironmental parts of BV. Furthermore, Ravel et al.'s study gave rise to a slew of subsequent studies, all of which took the racial/ethnic groups as given and began to place value judgments on "risky behaviors," "healthy vaginas," and "good" and "bad" vaginal microbiomes. These qualifications of vaginal microbiomes (and the vaginas they came from) correspond directly to race and/or ethnicity.[44] What appears to be microbiota differentiated simply by race is microbiota affected by what it means to be a Black or Hispanic woman in the United States.

Funded by the iHMP, Jennifer Fettweis has made strong claims connecting preterm birth in African American women with the specific microbial taxa in their vaginas.[45] In subsequent work, Ravel develops novel strategies to improve women's health by biologically modifying the microbiome.[46] In 2021, the Ravel Lab had two active NIH-funded studies, "Influence of Modifiable Factors on the Vaginal Microbiota and Preterm Birth" and "Revealing the Role of the Cervico-vaginal Microbiome in Spontaneous Preterm Birth." Millions of dollars are being used to fund studies of microbes in the vagina, trying to fix the problem of preterm birth by fixing a broken vaginal microbiome. These studies enact what sociologist Troy Duster has called the molecular reinscription of race[47] and anthropologist Duana Fullwiley's molecularization of race,[48] but on a microbial level. This science seeks probiotic interventions to prevent adverse outcomes while failing to account for the well-documented evidence that racial discrimination, chronic stress, and other disparities contribute to preterm birth.[49] Following this, the inclusion of perspectives from a social science expert in public and/or sexual health, epidemiology, or medical anthropology in vaginal microbiome studies is crucial to foregrounding the biosocial entanglements, such as structural racism, that can affect microbial populations in the vagina.

Global Guts

When I was in the Gordon Lab, one postdoctoral researcher had a project known colloquially around the lab as the global gut study: analyzing children in different areas of the world, looking at the establishment of their first gut microbial communities through the first three years of life. This postdoc was interested in the interplay of host genotype, environment (culture, diet, climate), and early childhood development (interaction with caregiver, transition from breast milk) with the gut microbiota of different geographical populations. Because she was comparing African, South American, and U.S. subjects, she asked, how does gut microbiota react to "Westernization"?[50] The global gut study (actually titled "Human Gut Microbiome Viewed across

Age and Geography") collected thousands of fecal samples, over years, from hundreds of people in different parts of the world, "to examine how gut microbiomes differ between human populations when viewed from the perspective of component microbial lineages, encoded metabolic functions, stage of postnatal development, and environmental exposures."[51] Yatsunenko struggled with the great variation in her data and was challenged to find group effects. In one lab meeting, she lamented, "Each sample is only representative of that one person on that one day. There are no features of significant association with malnutrition across all families. All families have to be looked at individually. There is little overlap of ages where kids became malnourished." She often reported that her comparisons using regression models weren't really working: "When samples don't cluster on a plot, you have to go back and look at specific samples to see what's going on—when you can't statistically explain your results. We can't dissociate diet changes, ready-to-use-therapeutic-food use, and malnutrition from age and the changes that correlate." Then she began using another approach, starting with normalized EC tables: comparing two groups of matched samples and calculating the EC ratio, she began to see how each microbiota was functioning differently. ECs are enzyme commission numbers, a numerical classification scheme for enzymes based on the specific chemical reactions they catalyze. ECs show what bacterial genes are involved in things like vitamin biosynthesis and metabolism. For human microbial ecologists, identifying ECs are more important than determining bacterial species because ECs show what is functionally going on in a microbiota. For example, Yatsunenko compared a twin with *kwashiorkor* and the healthy co-twin sampled at the same time. She began to report that she saw the phylogenic (evolutionary relationship between organisms) diversity of the microbiome increasing over time and that diversity seemed to stabilize when exclusive breastfeeding stopped. When she stopped looking for similarities and instead started looking at the diversity of microbiota, her results started to take shape. In the data, she found the effect of antibiotics and the introduction of new foods on the microbial populations. She published this work in *Nature,* and it radically impacted scientific understandings of how the microbiome is constituted in early life and how the microbiome differs in relationship to worldly populations. Yatsunenko's study established that children's microbiomes develop relatively the same everywhere and was the first study comparing the gut communities of humans living different kinds of lives in different places. The most important takeaways focused on how the similarity of microbiomes among family members extended across cultures and how the microbiome needs to be evaluated in relationship to nutritional needs and the impact of Westernization.

This paper had several significant and entangled outcomes regarding how microbial populations are established as a factor of age, geography/cultural traditions, and diet. Yatsunenko and her colleagues recognized that these elements are not separable from human lives, nor from what microbes dominate the gut and what functions they perform. Here the global gut study takes a different trajectory from the vaginal microbiome research. Whereas in the vaginal studies, racial/ethnic groups were the starting point of finding difference in microbes, in the global gut study, people were categorized by diet and lifestyle. More consequentially, socioenvironmental factors, not presumed race, were determined to be the most important to microbial constitutions. Diet, birth, and nursing practices; how many people slept in a house; and what they ate contributed to the makeup of microbiomes. The expanded inclusion of ethnographic data and social science approaches can broaden and diversify what matters to microbiomes, instead of replacing biological reductionism with social reductionism.

The subjects of the global gut study were divided into three "geographical" groups: Malawian, Amerindian, and U.S. The Malawians were identified through their nationality but were from four distinct rural communities; the Amerindians were ethnically Guahibo Indians living in two villages in Venezuela; and the U.S. subjects were from urban populations in the cities of Boulder, Philadelphia, and St. Louis. Yatsunenko found that the "Malawian" and "Amerindian" microbiomes were good at breaking down starches, whereas the "U.S." microbiome specialized in the degradation of glutamine and other amino acids.

This made sense from a digestion point of view, since these Malawians and Amerindians ate mostly corn and cassava, whereas subjects from the United States ate a lot of animal protein. Other research has shown that rather than designating human microbiomes as "Indigenous," "traditional," or "modernized," it would be more precise to divide into meat and plant eaters.[52] A global gut contains all these chemical particularities and microbial gene representations, which were the result of where people lived and, consequently, what they ate.

The interpersonal difference in microbes was much greater between children than between adults, showing that microbial populations are wildly diverse in the beginning of life and tend to stabilize and homogenize with age. While the microbes of babies in all three geographic areas underwent transformations through the first three years of life, the assemblages of bacterial species, as well as their functional genomics (what those bacteria did in the gut), were strikingly different in the U.S. samples. Malawian and Amerindian baby microbiomes had more genes for utilizing the readily available sugars in breast milk and for making vitamin B2, possibly because U.S. babies get more B2 from their mothers, who eat more meat and dairy. Although it appears that conventional categories of Global North and South contrast the "developed" American microbes against the "underdeveloped" African and Amerindian ones, reifying them through studies of microbes, human microbial ecology can contribute to a more complicated view of humans. Yatsunenko told me:

> I think it's too early to identify populations by their microbiomes. I guess you can do that, but the boundaries between populations probably would fluctuate and remain fuzzy given the lability of the microbiome due to dietary changes, along with exposure to antibiotics and disease. I'm not sure how much geography contributes to the identity of the microbiome because we have globalization.

Here the microbial ecosystem begins to displace worldly continents. For Yatsunenko and her colleagues, the ways of life that contribute to the establishment and alteration of the microbiome supersede physical location or conventional ideas of population. "Continent of origin" genomic hap-mapping tells very little of the story of diversity compared to human microbiomes, which conclude that there are more similarities by diet—and potentially food economies—than by "race." Yet it is complicated; this research still deals heavily in the currency of racial–national identities by designating microbiomes as U.S., Malawian, and Amerindian. And though the microbiome scientists with whom I worked view malnutrition as a complex, multifaceted problem with contributing economic, biological, genetic, and social factors, they still hope to treat it with probiotic intervention. Fixing a malnourished microbiota with food or bacterial supplements circumvents sociomaterial vulnerabilities, the very daily dilemmas of sewage, food, population density, and poverty that bring these microbiomes into being.

Microbiomes are amalgamations of practices of everyday life, the small details of human existence, with the evolutionary cohistories and genomes of humans and microbes. One of Dr. Gordon's favorite phrases to repeat around the lab was "we are human *doings,* not just human beings." The global gut study found that when people eat mostly plants, when they breastfeed for several years and live in small dwellings with lots of extended family members (and perhaps close to livestock), their gut microbes adapt to those very specific circumstances. Bodies (microbial, and therefore human) draw as much nutrition as they can from the food they have available. As the Malawian and Amerindian babies' diets shift from breast milk toward high-fiber plant foods, their gut bacteria become loaded with genes for breaking down more complex sugars and starches. The microbes that have ECs encoded for glutamate synthase and alpha-amylase thrive, and those skilled at vitamin biosynthesis (required of a more fat-rich diet) retreat, die off. Because the U.S. babies likely have a lifelong diet of refined sugars ahead of them, the bacteria with genes for harvesting these nutrients become more abundant, and those that can break down amino acids become plentiful, aligning with their hosts' high-protein diets. These are domains, small and big, in which the forms of individual and collective existence are at stake.

Biosocial Intersectionality

Physician and epidemiologist J. Dennis Fortenberry states that oppressive categories of race continue to manifest in medicine because of demands for social "inclusion" in biomedical studies.[53] As long as clinical research sees categories of race as preformed and rigid, and social/biological effects as separate and exclusive, the biomedical knowledge produced will incorporate these binary reifications.

In that vein, Helmreich asks, "Would it make sense to take a more sophisticated approach and ask how social categories like race and processes like racism—and its attendant stresses and deprivations (and, in some cases, privilege)—can reach into people's biologies and reshape their microbiomes?"[54] But he warns that to molecularize the environmental influences on race is as problematic as the molecularization of race itself. I suggest something different than reductive biologizing: interpret microbiomes as biosocial relationships in process rather than reinforcing the separation of biological and social influences. Warning against molecularizing the social implicitly affirms the distinction between a "social" and a "biological." Instead of insisting on a division between the biological body and its social environments, I follow Lock and Nguyen to examine the dynamic process of embodiment, "also informed by the body, itself contingent on evolutionary, environmental, social, economic, political and individual variables that have impinged on it over time and in space."[55] Instead of asking scientists to do qualitative analyses of the "social" categories and processes at work in the microbiome (for which they are not trained), what if social scientists brought critical expertise to microbiomic partnerships, helping to expand the parameters of what a microbiome is?

Helmreich is concerned about complicated social practices being reduced to simplistic environmental influences and worries that the "microbiome" is an object taken for granted, separate from the technologies that produce it. Certainly microbiome science has emerged as a direct result of next-generation shotgun sequencing, which has produced what we think of as microbiomes. Acknowledging that the technoscientific apparatuses, systems, and legacies of knowledge (including legacies of anthropological knowledge) make up "microbiomes" as much as the microbial organisms themselves, I suggest an ontological trajectory of feminist science studies that takes seriously the sociomaterial as one (multiple) object. New materialism and feminist STS are pushing social analysis to reevaluate relationships with science, nonhumans, and our own disciplinary commitments. These turns have faced criticisms; there is unease about the foreclosure of the political, the social, and the human. In a special issue of *Catalyst: Feminism, Theory, Technoscience* on "Feminism's Science," editors Banu Subramaniam and Angela Willey address these issues:

> We—feminist science studies scholars—must learn to think of ourselves not only as critics or students of science, but as makers of scientific knowledge. What sciences would we put on our proverbial boots to march for?[56]

Instead of enforcing the biological and social binary, this new thinking enacts Karen Barad's ethico-onto-epistem-ology. In *Meeting the Universe Halfway,* Barad points to the inseparability of ethics, ontology, and epistemology when engaging in knowledge production; that knowing is a material practice of engagement; and that boundary production between disciplines is itself materially discursive.[57] I see this enacted as microbiomes coming into being through biosocial relationships across disciplines, across microbial and human bodies.

To attempt this kind of intercession, my work in Bangladesh contributed ethnographic data that corresponded to the biological samples. My observations on ways of living, caring for children, and feeding families tried to help evolve the scientific view of study subjects. I tried to "un-North" the microbiome by having conversations with study subjects about their experiences, with careful attention to how responsibility and compliance are assigned to the most vulnerable actors, attempting to reroute scientific health interventions. During my time in Bangladesh, I spoke primarily with mothers, but I also interviewed other extended family and household members about being enrolled in the study, how they understood microbes, and what they cared about most. It was unprecedented for the lab scientists with whom I worked to hear the voices of their subjects and to learn how they felt being part of the study and what was at stake for them. They had never confronted the qualitative face of malnutrition or seen photos of the mothers and children and what their homes and communities looked like. When I asked Gordon Lab members to ask themselves, why these communities? why these bodies? I was trying to unghost race by making racial and ethnic thinking explicit. Interviewing Bangladeshi scientists and researchers was also an important part of this story—as there is increasing recognition that global research collaborations rarely demonstrate equity in publishing, coauthorship, or funding between resource-rich, or Global North, countries with partners in middle- to low-income, or Global South, countries.[58]

For microbiome scientists, a biosocial partnership emphasizes that relationships between and beyond humans need to be considered in designing and interpreting observational and interventional microbiome studies. And for social scientists, a focus on human–microbe relationships may push us to "reimagine health and well-being as more than human concerns,"[59] perhaps even reorienting an analysis of race that goes beyond fixed, stable categories of biology, culture, and human.[60] Feminist scholars give us a good starting place: what Yates-Doerr calls *careful equivocation,*[61] attuning ourselves to disparate bio/social binaries but working together to shape global health imperatives, or Elizabeth Roberts's *bioethnography,* "a method for combining ethnographic and biological data to arrive at better understandings of the transmitted life circumstances that shape health and inequality."[62]

Research like the global gut study creates an opportunity to use ethnography as a "crucial methodological tool for achieving better comprehension of health services at all levels of analysis."[63] A microbe-expanded view of humans will provide space for social scientists and life scientists to consider how social environment and biological bodies work in inextricable consort. Fortenberry claims that "microbiome research needs ways of thinking about human diversity to match its burgeoning sophistication with microbial diversity."[64] Making this possible will necessitate partnerships, where social scientists are active participants in deconstructing and reformulating categorizations of people, while accounting for the biosocial effects of living within the structural violence of racism. Race is meaningful as a category if the designation can be seen as an ethico-onto-epistemological one[65] and not biologically a priori: race as the sum, not the addend. Poverty, resources, social inequities, and race are not peripheral to how the microbiome is studied or understood, nor are they singularly explanatory. Race as a process and all its attendant biosocial outcomes are important to study along with microbiomes, especially as growing knowledge shows that diet, toxic exposures, housing, and health care access affect what microbial populations we have.

Social uprisings centered around racial injustice and a long-overdue reckoning with the racism embedded in academia and science are perhaps moving the conversation in the direction of social equity. Groups of grant makers, scientific researchers, and scholars are writing and speaking on essential changes that need to be made in genomic and microbiome sciences. The Black Microbiologists Association was created in response to the sustained underrepresentation of Black people in microbiological education, training, and research and aims to advocate for Black microbiologists and to pursue equity in academia, industry, and government.[66] In fall 2021, a group out of the NIH National Human Genome Research Institute wrote a paper titled "Cultivating Diversity as an Ethos with an Anti-racism Approach in the Scientific Enterprise."[67] In promoting diverse scientific team building, the authors reflect on how structural racism impedes representation in genomic science and, consequently, limits scientific innovation. Rather than diversity as a performative gesture, they call for diversity as an ethos in genomics, requiring institutional change and intentional antiracism, asking scientists to interrogate both systems and personal perspectives: "(1) How are studies being conducted in this lab? (2) Who are the scientists conducting these studies? and (3) How does the research environment cultivate space for reflection across the research process, individually and collectively?"[68]

My post–Gordon Lab partnerships have both broadened in scope (large research groups across many institutions and disciplines) and narrowed in on topics that center social equity, justice, and antiracism. In many ways, I have turned toward junior microbiome scientists, who see social conditions, care, and equity as critical to their research. Rather than struggling against the peripherality of anthropology in microbiome science, these new collaborations center the biosocial. The Microbiome and Social Equity Working Group (of which I am a founding member) started in 2019 by Sue Ishaq, an animal and veterinary microbiome scientist at the University of Maine, has the mandate of education and training: diversifying research topics and methods and advocating for evidence-based public policy surrounding microbiomes. Across many scientific, social scientific, public, and environmental health fields, we published a white paper in *mSystems* in summer 2021, tying together the biology of microbiota with socioeconomic imperatives:

> Yet engagement with beneficial microbiomes is dictated by access to public resources, such as nutritious food, clean water and air, safe shelter, social interactions, and effective medicine. In this way, microbiomes have sociopolitical contexts that must be considered. The Microbes and Social Equity (MSE) Working Group connects microbiology with social equity research, education, policy, and practice to understand the interplay of microorganisms, individuals, societies, and ecosystems.[69]

This group also ran a symposium of speakers who ranged from anthropologists to molecular biologists, neuroscientists, and psychologists, talking about microbes and social equity topics around homelessness and environmental microbiomes, equity in infectious disease intervention, and Indigenous narratives of microbes.[70] The journal *mSystems*, the flagship journal of the American Society for Microbiology, ran a special issue in fall 2021 titled "Deciphering the Microbiome." In that issue, a group of mostly junior scientists out of Genomics and Bioinformatics Research in the U.S. Department of Agriculture's Agricultural Research Service argue ardently that microbiome sampling thus far has been extremely limited in its representation of global populations and that the science itself has historically excluded people because of race, ethnicity, or location in low- or middle-income countries.[71] The authors propose that researchers are needed who understand the logistical and social needs of study populations and that equitable collaboration, mentoring, and training are needed if microbiome science is to successfully implement solutions to global health problems:

> Shifting the research culture in microbiome science by making research and training more inclusive will improve retention of underserved scientists and contribute to the field's intellectual growth. Such endeavors are opportunities to broaden our perspectives of how the environment, social identity, biopsychosocial factors, and cultural practices affect microbiome science around the world but are also imperative to restore equity and social justice in the field.[72]

These groups of researchers are taking microbiome research to task, to correct the course of racism in both the ethics and the equity of the production of microbiome knowledge, in who serves as research subjects and how. In that same issue, I published "Chasing Ghosts: Race, Racism, and the Future of Microbiome Research,"[73] a collaborative paper with Travis De Wolfe, Mohammed Rafi Arefin, and María Rebolleda Gómez (a microbiologist, a geographer, and an evolutionary ecologist, respectively) about the pressing need for an antiracist move in microbiome science. Building on the idea of ghost variables that I develop here and elsewhere,[74] "Chasing Ghosts" advocates that microbiome science cannot continue to "explicitly and implicitly deploy the variable of race in research without accounting for the relevant forms of racism that impact people and environments."[75] We review racist histories of microbiome science and call scientists to action to interrogate the unreflective uses of racial categories in their research. The paper concludes with a three-part, actionable plan for scientists to build an antiracist microbiomic future:

> First, institutional changes must be made in funding, publishing, hiring, and recruiting practices. Second, transdisciplinary collaboration across the biological and social sciences must be established as essential and customary. Third, study populations and BIPOC communities must be engaged in the research and empowered through the science.[76]

Instead of viewing differences in microbiomes as a consequence of race alone, we assert that microbiome disparity is caused by structural drivers (that affect some races disproportionately). We contend that replacing ghost variables with an analysis of structural drivers and their associative factors results in a more rigorous and just microbiome science. Structural drivers are often excluded from the purview of biomedical and biological research as outside the scope of scientific methods or questions. So, to be antiracist, microbiome science must be transdisciplinary. What is needed to address the interacting systems of microbes, health, and environment exceeds the expertise of individual disciplines. This work, written for a scientific audience, and imminently impactful, has proved to be an extremely productive and worthwhile move toward an anthropology of microbes.

It is not the goal of these investigations to arrive at one definitive idea of race or to concretize race as a social or biological thing to be grasped. The problems now lie in the way in which race is used to create scientific knowledge about the microbiome. Imagining microbiome science as postracial only causes more harm. It will take the careful cooperation of those in the social and biological sciences to navigate the labyrinthine biosociality of human–microbe relationships to make interventions that matter, to chase the ghosts of race into the light.

Conclusion

When You Give a Mouse a Microbe

I'm really quite bad at conclusions. Maybe because the work doesn't ever feel done, just like it never felt continuous or clear, just a fragmented amalgamation of experiences. I didn't take many pictures in the Gordon Lab. Maybe five total—the lab, benches empty in the early morning. Unpeopled equipment, the mouse facility, the outside of the building. It was hard to take pictures, like it is still hard to tell ethnographic stories of my time in the lab. Partly because the lab members were hyperaware of their participation in my research. Some explicitly expressed their concern and discomfort—they didn't want to be subjects and said so. Dr. Gordon was giving me a paycheck and health insurance, feeding my family, and providing us with many kinds of care. More than that, he was bestowing upon me a once-in-a-lifetime opportunity—he was making my career. How could I ask him to pose for photos? It seemed so intrusive and wrong. The part where we were collaborators and partners and the part where lab members were the objects of my study did not jibe; they were incommensurable. Conversely, I have hundreds of photos of Dhaka, of the research labs, nutrition centers, homes, streets, markets. I have so many pictures of Bangladeshi faces and bodies and none of Gordon Lab members. In hindsight, it is easy to see my whiteness and class privilege at work, the ways in which I very easily and without question "ethnographized" the people I met and with whom I worked in Bangladesh, but was too uncomfortable to ask members of the lab to be subjected to the same invasions. The limitations of my ethnography in Dhaka, the things I didn't and couldn't ask, were very rigidly bound. I was "part of the lab family," a family membership that required loyalty. I walked too near to narratives of damage,[1] and that is troubling and embarrassing to me now. I wish I had held the Gordon Lab and myself more accountable.

As I prepared to leave the lab and St. Louis, I made this last entry in my fieldnotes:

> After being in the lab for over a year, we're finally leaving. Trying to really make plans for the future of this collaboration—maybe securing funding for further trips, research, work. This is, of course, both good and bad. Part of me really really needs to break from JG. I realized yesterday when Joe and I were talking that my fieldwork year really wasn't typical. You don't often work at the place you are studying, and I realized we really did *work* this past year. We earned the money JG paid us. We wrote for the lab, we researched for the lab. Which is fine—good actually, because I don't feel as bad about the money, we earned it. But it also means I've been in a very specific mindset, that I need to transition out of to start thinking about my own work. I am and always will be so grateful to him. And that gratitude will always be tangled up in frustration and struggles for independence and misunderstandings.

After finishing and defending my dissertation in 2014, I began to write about my Gordon Lab work for various journals. I submitted a paper to *BioSocieties* and, when I got reviews back, sent the paper to Dr. Gordon. I sent it out of respect, for his opinion and our partnership. It was to let him know what I was doing; I wasn't seeking his approval. He returned it to me with extensive comments and edits. He deleted all speculative language, changed anything that seemed to deviate from the standard lab literature and public-facing lab doctrine. It took me a few months to figure out how to respond. I felt genuinely hurt, by the authority he claimed in editing my anthropological work and by how constricting I suddenly felt our partnership was. In October 2015, Dr. Gordon wrote me an email, the last communication we would have after several years of a collaborative relationship. And that's how it ended. I couldn't make the changes he wanted me to make, and he couldn't sign off on the paper as it was. I share this here, not to sensationalize or criticize, not to garner sympathy from the social science crowd who are anxiously waiting with "I-told-you-sos," but to show how very complicated this all is.

Dr. Gordon thought there were inaccuracies and misinterpretations in my paper, and it is likely he will feel the same about my account in this book. In Dr. Gordon's edits of my paper, it is clear how he wanted me to talk about the lab's work. It is work that means so much to him, to which he is so dedicated. I never doubted for a second that he actually did want the best for me, cared about my career development, and believed with his whole heart that his work was for good. And isn't it? But somehow there isn't room in the sentence—for something like gratitude and humility to the Gates Foundation for their support but to also question what it means for the science to be influenced by such a mighty funding entity. There didn't seem to be room in the work to appreciate Dr. Gordon and believe in the science, and also see the cracks. How was I supposed to talk about this? To turn my experience into job talks and articles and book chapters? There are things about my time in the Gordon Lab that I will never talk about, things I never want to talk about. I won't write about everything, not because it is scandalous or outrageous, but because my entire relationship with the lab and Dr. Gordon was balanced precariously on a promise of trust. This trust was personal, and also trust in a different kind of disciplinary data that you might not understand.[2] I value and respect the people with whom I worked and have no desire to bring them any harm or discomfort, professional or otherwise. And in the back of my mind I know—how could I ever convince another scientist into a partnership if I've somehow burned my last collaborator? Why would anyone want to work with me?

And what about saving lives? What about the white male egoist savior? How do we define the category of the powerful flexing of their muscle under the guise of help and care? I wonder if one can find instances of anthropology saving lives. We do have a lot to say about how others go about it and where they go wrong. I feel like instead of doing the work, we prefer to problematize the idea of saving—whose idea is it, and what does it look and feel like when you are the one being "saved"? This is extremely important work, but does it *do* enough? I return again to Paul Farmer, as he implores anthropologists to practice an *anthropology of structural violence,* which "necessarily draws on history and biology, just as it necessarily draws on political economy. To tally body counts correctly requires epidemiology, forensic and clinical medicine, and demography."[3]

Often when I give talks about the frictions of interdisciplinarity, I am asked by audiences, what did I finally decide about the Gordon Lab work? I always say that all the kids in the Microbiome Discovery Project lived. I say that if nothing else, I want to be on the side of living children. I find Farmer's formulation of an anthropology of structural violence so helpful when wrestling with this. From the purview of the Gordon Lab, I felt compelled to tally the living, not the dead. This says something.

> An anthropology that tallies the body count must of course look at the dead and those left for dead. Such inquiry seeks to understand how suffering is muted or elided altogether. It explores the complicity necessary to erase history and cover up the clear links between the dead and near-dead and those who are the winners in the struggle for survival. Bringing these links—whether termed social, biologic, or symbolic—into view is a key task for an anthropology of structural violence. I will argue here that keeping the material in focus is one way to avoid undue romanticism in accomplishing this task. An honest account of who wins, who loses, and what weapons are used is an important safeguard against the romantic illusions of those who, like us, are usually shielded from the sharp edges of structural violence. I find it helpful to think of the "materiality of the social," a term that underlines my conviction that social life in general and structural violence in particular will not be understood without a deeply materialist approach to whatever

> surfaces in the participant-observer's field of vision—the ethnographically visible.[4]

This speaks powerfully to me, obligating anthropologists to use ethnography not just to theorize about systems of inequity and violence but to see what is actually happening to the bodies we write about. Malnutrition is not just complexly colonial, political, and historical; it also acts. It kills people. Is there any way to do this work "right"? I don't have an answer. Is there an answer? In the interceding years, I have presented this work to the delight and horror of anthropologists, geographers, public health audiences, students across disciplines, STS scholars, and microbiologists. I got really good at punching up the potentialities of collaboration and, in later years, finally confessing to the failures. Anthropologists who hear and read my work are obsessed with the shittiest parts, they ask questions upon questions about the miscommunications and power struggles, they want to hear about how science plays out exactly as they expected it to. Maybe even in spite of themselves, they *wanted* the collaboration to fail, because it confirms everything we think we know about translational medicine and technoscience. It allows for a haughty and self-righteous that's-what-you-get-for-trusting-science.

So, for years after my fieldwork, I struggled. I will always believe that Dr. Gordon is a truly exceptional scientist and human being. I believe unequivocally that he cared for me. He distributed so much joy; he instilled it in his research and inspired it in those who worked with him. He brought me into the folds of his lab, taught me about his work, showed me his vision of how he wanted to use science to make the world better. But he was also wrong about some things. His narrow view that microbiome intervention is the cure for malnutrition made it impossible to see science as part of the systems of power that caused the problem in the first place. I was desperate to follow Callard and Fitzgerald—"we want, in other words, as interdisciplinary scholars, not to seek mythical platforms for equal exchange, but to keep learning different ways of being unsettled together"[5]—but couldn't find my way.

Collaborating was the greatest thing I could have hoped for in my research. At this point I can also say that collaborating was traumatic. Because the value of my fieldwork (within my own discipline, to peer reviewers and grant makers) seemed to be conspicuously tied to my very unique access to Dr. Gordon and his lab and the loss of that link felt like it rendered my work meaningless. The actual break with Dr. Gordon caused real, emotional pain. I felt I'd let him down, failed at creating a durable partnership, wasn't a good enough anthropologist to make it work. But I also felt angry at what I felt was his unwillingness to let me do the work I wanted to do, to ask the questions I wanted answers to. I was confused that he disapproved of my work and cut me off, that his vision for our work together was so different than mine, and that there was no room for me in his view. It felt like all his generous encouragement turned to criticism—I got the science wrong, I got the anthropology wrong.

In the end, Dr. Gordon believed we had very different views of how a collaborative partnership should operate, and he was absolutely right. A rotating cast of infectious disease MDs, public health grad students, and other non–social scientists have served the role of "societal issues" expert in the Gordon Lab microbiome work since. I have saved several voicemail messages on my phone—going back twelve years now—just to prove to myself sometimes that this collaboration did happen. He *did* like me enough to leave warm, exuberant phone messages to me. Periodically, I am overcome with a desire to write or call Dr. Gordon, to reestablish contact, to reenter the glowing, warm orbit of his good graces. Sometimes I really do want to keep working with him, to make an anthropology of microbes a reality.

I think that with collaboration, shit is inevitable. There is no way around it, no way to remove or avoid it. The only way is through, and I'm not even sure there's a way out. A friend asked me recently what I would work on next, once this book was done. I replied, "More microbiome, of course." "Why?" she exclaimed, legitimately shocked. The subsequent collaborations I have been lucky to be part of are quite different from my partnership with Dr. Gordon. To a great extent, my collaborators are junior faculty and researchers: women, queer, and gender-diverse scholars; scientists of color; people who have deep investments in transforming science as they build labs and careers of their own. They are largely committed to figuring out how to do science with anticolonial, antiracist, and feminist questions, practices, and interventions at the center. They are getting things done. When I started this project, I thought anthropology would be a peripheral, observational node to microbiome science. Working with Dr. Gordon helped me develop meaningful practices of ethnography and figure out what anthropology was for me. I didn't anticipate that microbiome science would also change, that a new generation of scientists would ask more of their disciplines and transform them in kind. I will keep working on the microbiome because I still believe that anthropology and human microbial ecology are interacting, interdisciplinary, coevolving messmates and kin. I still believe this work is important—and possible.

Acknowledgments

When your project spans thirteen years, a postdoc, the job market, and two kids, and you don't feel like you've been particularly good at keeping track of everything, you sort of dread having to write acknowledgments for fear of whom you will leave out. I am deeply indebted to people at both unfixed ends of my research network, without whom this project would not exist. Endless thanks to the mothers in Mirpur who let me into their homes and talked to me about their babies, to the field research assistants who taught me how the work was done, and to Israt for making sense of things. Tahmeed Ahmed, Shamshir Ahmed, and Sayeeda Huq helped navigate the icddr,b, provided invaluable insight and support, and made sure I was always fed. The members of the Gordon Lab kindly accepted me into their workspaces and lives and were smart, thoughtful, and patient with all the questions I asked them. Particular thanks go to Philip Ahern, Vanessa Ridaura, Vitas Wagner, Tanya Yatsunenko, and Michelle Smith.

A million years ago, at the New School for Social Research, Hugh Raffles was an exceptional writing mentor and an overall outstanding advisor and let Bea pull books off the shelves in his office when she was a toddler. I can't thank him enough for all that, and for the letters upon letters upon letters that he wrote. Nicolas Langlitz provided crucial perspective and was an excellent editor. BALD (Brie Gettleson, Leilah Veviana, Diego Caguenas) was the PhD cohort with the mostest, and I am honored to have studied with them.

At NYU Emily Martin taught me to always stay curious and brought me to microbes. Helena Hansen was the most compassionate, kind, and profoundly smart postdoc advisor and mentor I could have hoped for. Rayna Rapp, who is truly the GOAT, reminded me about where my heart and mind lie in this work, unfailingly encouraged me, and helped keep feminist theory at the forefront. She continues to be the kind of scholar and person I strive to be. When I finally got a tenure-track job, I was so lucky to have Sam Muka and Alex Wellerstein as colleagues who made me feel so welcome at Stevens.

At Washington University, Jeff Gordon provided a space to grow intellectually while also letting me work to enact real-world change. I will always be tremendously grateful that I got to learn from him and for the collaborative opportunity he gave me. The joy and care he brings to science and gives to those around him make me so glad that he exists. I apologize to him for using the word *shit* so much in this book.

David Bond, Katayoun Chamany, Alison Cool, Kadija Ferryman, Beck Jordan-Young, Katrina Karkazis, Martine Lappé, Theresa MacPhail, Alex Nading, Dana Powell, and Liz Roberts are all generous and brilliant scholars and humans. Emily Yates-Doerr gave me (and continues to give me!) wisdom and care I would be lost without. Janelle Lamoreaux shows me how to be a better thinker and friend, and I have relied on her for countless forms of support.

Jason Weidemann at the University of Minnesota Press was a good and calm editor who had lots of cups of coffee and beers with me in many different cities. I also thank Stefan Helmreich for deeply thoughtful reads, and I thank those anonymous reviewers, some of whom were tough and some of whom cared about making the book the best it could be. Thank you to Explosions in the Sky for providing a loud and motivating soundtrack to write by for years and years. Thanks to my mom for always supporting my work, and thanks to Jim for always being interested. Thank you to my sister Crystal for being my true family.

My deepest thanks are saved for Beatrice, Lydia, and Joe. Beatrice showed me how truly intertwined biology and love can be. She stuck it out all those long, long days that "mama typies"; learned how to say "dissertation" at twenty months old; and grew into a genius, kind, curious ten-year-old who understands what microbiomes are. Liddy let me leave her at five weeks old to adjunct, to postdoc, and still she shined like the brightest sunbeam. She became a keen and unique person who surprises me every day and who came up with many title options for this book. Joe took all the important trips with me, to St. Louis and to Dhaka; read a million versions of a million things; and talked about microbes even when we were exhausted at the end of the day and all we really wanted to do was watch TV. I thank him for always making me be my best self and for making my life one worth living well.

Research for this project was supported by grants from the National Science Foundation (grant 1027035) and the New School for Social Research, as well as by resources and funds from the Gordon Lab at Washington University and from icddr,b.

Notes

Preface

1. Emily Yates-Doerr, "Antihero Care: On Fieldwork and Anthropology," *Anthropology and Humanism* 45 (2020): 233–44, https://doi.org/10.1111/anhu.12300.
2. Jody A. Roberts, "Reflections of an Unrepentant Plastiphobe: Plasticity and the STS Life," *Science as Culture* 19, no. 1 (2010): 103, 101–20, https://doi.org/10.1080/09505430903557916.
3. Rosanne Cecil, ed., *Anthropology of Pregnancy Loss: Comparative Studies in Miscarriage, Stillbirth and Neo-natal Death* (Oxford: Berg, 1996); Linda L. Layne, *Motherhood Lost: A Feminist Account of Pregnancy Loss in America* (New York: Routledge, 2003).
4. Maria G. Cattell and Marjorie M. Schweitzer, eds., *Women in Anthropology: Autobiographical Narratives and Social History* (New York: Routledge, 2016); Carrie Friese, "When Research Bleeds into Real Life: Studying Reproductive Ageing While Ageing Reproductively," *Somatosphere* (blog), November 2015, http://somatosphere.net/2015/11/when-research-bleeds-into-real-life-studying-reproductive-ageing-while-ageing-reproductively.html; Rayna Rapp, with Faye Ginsburg, "Fetal Reflections: Confessions of Two Feminist Anthropologists as Mutual Informants," in *Fetal Positions, Feminist Practices,* edited by Lynn Morgan and Meredith Michaels, 279–95 (Philadelphia: University of Pennsylvania Press, 1999); Rapp, *Testing Women, Testing the Fetus: The Social Impact of Amniocentesis in America* (New York: Routledge, 1999).
5. Rapp, *Testing Women, Testing the Fetus,* 121.
6. I thank all the Society of Cultural Anthropology "Displacements" panel mother antiheroes, but especially Emily Yates-Doerr, for making me think

academically about motherhood and how to turn silences into shouts or quiet but powerful statements.

7. Maya J. Berry, Claudia Chávez Argüelles, Shanya Cordis, Sarah Ihmoud, and Elizabeth Velásquez Estrada, "Toward a Fugitive Anthropology: Gender, Race, and Violence in the Field," *Cultural Anthropology* 32, no. 4 (2017): 537–65; Rebecca Hanson and Patricia Richards, "Sexual Harassment and the Construction of Ethnographic Knowledge," *Sociological Forum* 32, no. 3 (2017), https://doi.org/10.1111/socf.12350; Sinah Theresa Kloß, "Sexual(ized) Harassment and Ethnographic Fieldwork: A Silenced Aspect of Social Research," *Ethnography* 18, no. 3 (2017): 396–414; Robin G. Nelson, Julienne N. Rutherford, Katie Hinde, and Kathryn B. H. Clancy, "Signaling Safety: Characterizing Fieldwork Experiences and Their Implications for Career Trajectories," *American Anthropologist* 119 (2017): 710–22; Bianca C. Williams, "'Don't Ride the Bus!' and Other Warnings Women Anthropologists Are Given during Fieldwork," *Transforming Anthropology* 17, no. 2 (2009): 155–58.
8. At the time of my final manuscript edits, academic anthropology in the United States was in full tumult following the exposure of systematic and decades-long sexual abuse of students by Harvard anthropologist John Comaroff. Isabella B. Cho and Ariel H. Kim, "Lawsuit Alleges Harvard Ignored Sexual Harassment Complaints against Prof. John Comaroff for Years," *Harvard Crimson*, February 9, 2022, https://www.thecrimson.com/article/2022/2/9/comaroff-lawsuit. Of course this wasn't news or isolated to Harvard, and many senior anthropology faculty reacted with outspoken, knee-jerk defense of power by power. "Open Letter against Harvard's Treatment of John Comaroff," *Chronicle of Higher Education*, February 3, 2022, https://www.chronicle.com/blogs/letters/open-letter-against-harvards-treatment-of-john-comaroff. But at least people were starting to talk about it.
9. David Bond, "Contamination in Theory and Protest," *American Ethnologist*, November 2021, https://doi.org/10.1111/amet.13035.
10. Black feminist and Indigenous anthropologists TallBear, McLaurin and Bolles, Tuck and Yang, and Simpson have opened dialogues of representational refusal in significant ways.
11. Friese, "When Research Bleeds into Real Life," 5.
12. Michelle Téllez, "Why We Must Write: A Reflection on Tenure Denial and Coloring between the Lines," *Feminist Wire*, Summer 2016, https://www.thefeministwire.com/2016/07/michelle-tellez/.
13. Julie Johnson Searcy and Angela N. Castañeda, "Making Space for Mothering: Collaboration as Feminist Practice," Member Voices, *Fieldsights*, February 2020, https://culanth.org/fieldsights/making-space-for-mothering-collaboration-as-feminist-practice.

Introduction

1. Gabriele Berg, Daria Rybakova, Doreen Fischer, Tomislav Cernava, Marie-Christine Champomier Vergès, Trevor Charles, Xiaoyulong Chen et al., "Microbiome Definition Re-visited: Old Concepts and New Challenges," *Microbiome* 8 (2020): Article 103.
2. In human microbial ecology (and following that field, here), *gut* refers to the gastrointestinal tract, including all the digestive organs from esophagus to anus.
3. Julia Evangelou Strait, "The Father of the Microbiome," *Washington Magazine*, March 3, 2017, https://source.wustl.edu/2017/03/the-father-of-the-microbiome/.
4. Amber Benezra, Joseph DeStefano, and Jeffrey I. Gordon, "Anthropology of Microbes," *Proceedings of the National Academy of Sciences of the United States of America* 109, no. 17 (2012): 6378–81.
5. Maurizio Meloni, John Cromby, Des Fitzgerald, and Stephanie Lloyd, *Palgrave Handbook of Biology and Society* (London: Palgrave Macmillan, 2018), 5.
6. Elizabeth Reddy, "What Does It Mean to Do Anthropology in the Anthropocene?," *Platypus* (blog), April 2014, https://blog.castac.org/2014/04/what-does-it-mean-to-do-anthropology-in-the-anthropocene/.
7. Zoe Todd, "Relationships," Theorizing the Contemporary, *Fieldsights*, January 21, 2016, https://culanth.org/fieldsights/relationships.
8. Kathryn Yusoff, *One Billion Black Anthropocenes or None* (Minneapolis: University of Minnesota Press, 2018).
9. Hannah Landecker, "Antibiotic Resistance and the Biology of History," *Body and Society* 22, no. 4 (2016): 19–52.
10. Karen Brodkin, Sandra Morgen, and Janis Hutchinson, "Anthropology as White Public Space?," *American Anthropologist* 113 (2011): 545–56.
11. Association of Black Anthropologists, "ABA Statement against Police Violence and Anti-Black Racism," 2020, http://aba.americananthro.org/aba-statement-against-police-violence-and-anti-black-racism-3/.
12. Leanne Betasamosake Simpson, "Not Murdered, Not Missing: Rebelling against Colonial Gender Violence," in *Burn It Down*, edited by B. Fahs, 314–20 (Brooklyn, N.Y.: Verso Press, 2020); Jinthana Haritaworn, "Decolonizing the Non/Human," *GLQ: Journal of Lesbian and Gay Studies* 21, no. 2–3 (2015): 209–48; Zakiyyah Iman Jackson, "Animal: New Directions in the Theorization of Race and Posthumanism," *Feminist Studies* 39, no. 3 (2013): 669–85; Julie Livingston and Jasbir K. Puar, "Interspecies," *Social Text* 29, no. 1 (2011): 3–14; Sandra Styres, "Literacies of Land: Decolonizing Narratives, Storying, and Literature," in *Indigenizing and Decolonizing Studies in Education: Mapping the Long View*, edited by Linda Tuhiwai Smith, Eve Tuck, and K. Wayne Yang, 24–37 (New York: Routledge, 2018); Juanita Sundberg, "Decolonizing Posthumanist Geographies," *Cultural Geographies* 21, no. 1 (2014): 33–47; Kim TallBear, "Why Interspecies Thinking Needs Indigenous

Standpoints," *Fieldsights*, November 18, 2011, https://culanth.org/fieldsights/why-interspecies-thinking-needs-indigenous-standpoints.

13. Sarah Hunt, "Ontologies of Indigeneity: The Politics of Embodying a Concept," *Cultural Geographies* 21, no. 1 (2014): 27–32; Zoe Todd, "An Indigenous Feminist's Take on the Ontological Turn: 'Ontology' Is Just Another Word for Colonialism," *Journal of Historical Sociology* 29, no. 1 (2016): 4–22.
14. NIH Human Microbiome Project, https://hmpdacc.org/.
15. These were the core donors in 2010–11. The DFID has been replaced by the Foreign, Commonwealth & Development Office, and CIDA is now part of Global Affairs Canada.
16. Bill and Melinda Gates Foundation, "Bill and Melinda Gates Foundation Enteric and Diarrheal Diseases Strategy Overview," http://www.gatesfoundation.org/What-We-Do/Global-Health/Enteric-and-Diarrheal-Diseases.
17. PCOA, or principal coordinate analysis, is a method of multidimensional scaling often used by scientists studying the human microbiome. Using an input matrix to show dissimilarities between pairs, a PCOA plot usually shows how microbial communities from different donors or different body sites "cluster" together on a plot.
18. Felicity Callard and Des Fitzgerald's renovation of the tired concept of "collaboration" in *Rethinking Interdisciplinarity across the Social and Neurosciences* (London: Palgrave Macmillan, 2015).
19. Elizabeth F. S. Roberts and Carlos Sanz, "Bioethnography: A How-To Guide for the 21st Century," in Meloni et al., *Palgrave Handbook of Biology and Society*, 749; Emily Yates-Doerr, "Whose Global, Which Health? Unsettling Collaboration with Careful Equivocation," *American Anthropologist* 121, no. 2 (2019): 297–310.
20. Tim Ingold, "That's Enough about Ethnography!," *Hau: Journal of Ethnographic Theory* 4, no. 1 (2014): 383–95.
21. Heather Paxson and Stefan Helmreich, "The Perils and Promises of Microbial Abundance: Novel Natures and Model Ecosystems, from Artisanal Cheese to Alien Seas," *Social Studies of Science* 44, no. 2 (2014): 165–93; Stefan Helmreich, *Sounding the Limits of Life: Essays in the Anthropology of Biology and Beyond* (Princeton, N.J.: Princeton University Press, 2015); Alex Nading, "Evidentiary Symbiosis: On Paraethnography in Human–Microbe Relations," *Science as Culture* 25, no. 4 (2016): 560–81.
22. Paxson and Helmreich, "Perils and Promises," 185.
23. Helmreich, *Sounding the Limits of Life*, 65.
24. John Hartigan, "Social Theory for Nonhumans," https://manifold.umn.edu/projects/social-theory-for-nonhumans; Myra Hird, *The Origins of Sociable Life: Evolution after Science Studies* (New York: Palgrave Macmillan, 2009); Gísli Pálsson, "Ensembles of Biosocial Relations," in *Biosocial*

Becomings: Integrating Social and Biological Anthropology, edited by Tim Ingold and Gísli Pálsson, 22–41 (Cambridge: Cambridge University Press, 2013).

25. Elizabeth A. Wilson, *Gut Feminism* (Durham, N.C.: Duke University Press, 2015), 12.
26. Heather Paxson, "Post-Pasteurian Cultures: The Microbiopolitics of Raw-Milk Cheese in the United States," *Cultural Anthropology* 23, no. 1 (2008): 15–47.
27. Max Liboiron, *Pollution Is Colonialism* (Durham, N.C.: Duke University Press, 2021).
28. Many thanks to Stefan Helmreich for pointing this out and for helping me to articulate my position.
29. Charlotte Brives, Matthäus Rest, and Salla Sariola, eds., *With Microbes* (Manchester, U.K.: Mattering Press, 2021); Astrid Schrader, "Marine Microbiopolitics: Haunted Microbes before the Law," in *Blue Legalities: The Life and Laws of the Sea*, edited by Irus Braverman and Elizabeth R. Johnson, 255–74 (Durham, N.C.: Duke University Press, 2020).
30. Ståle Wig, "'Divorce Your Theory': A Conversation with Paul Farmer," *Savage Minds*, February 14, 2014, https://savageminds.org/2014/02/14/divorce-your-theory-a-conversation-with-paul-farmer-part-one/.
31. Kacy Greenhalgh, Kristen M. Meyer, Kjersti M. Aagaard, and Paul Wilmes, "The Human Gut Microbiome in Health: Establishment and Resilience of Microbiota over a Lifetime," *Environmental Microbiology* 18, no. 7 (2016): 2103–16.
32. Benezra et al., "Anthropology of Microbes," 6380.
33. Erin Koch, "Local Microbiologies of Tuberculosis: Insights from the Republic of Georgia," *Medical Anthropology* 30, no. 1 (2011): 81–101.
34. Jo Handelsman, "Metagenomics and Microbial Communities," in *Encyclopedia of Life Sciences* (Chichester, U.K.: John Wiley, 2007).
35. Sabrina Leonelli, "Global Data for Local Science: Assessing the Scale of Data Infrastructures in Biological and Biomedical Research," *BioSocieties* 8, no. 4 (2013): 449–65; Leonelli, "What Difference Does Quantity Make? On the Epistemology of Big Data in Biology," *Big Data and Society* 1, no. 1 (2014): 1–11.
36. Elizabeth Stulberg, Deborah Fravel, Lita M. Proctor, David M. Murray, Jonathan LoTempio, Linda Chrisey, Jay Garland et al., "An Assessment of US Microbiome Research," *Nature Microbiology* 1, no. 1 (2016): Article 15015.
37. Microbes as pathogens have long been an ethnographic object, and the contextualization and eradication of infectious diseases like tuberculosis, malaria, and HIV profoundly concern social scientists. Paul Farmer, *Infections and Inequalities: The Modern Plagues* (Berkeley: University of California Press, 2001); Didier Fassin, *When Bodies Remember: Experiences and Politics*

of AIDS in South Africa (Berkeley: University of California Press, 2007); Marcia Inhorn and R. A. Hahn, eds., *Anthropology and Public Health: Bridging Differences in Culture and Society* (Oxford: Oxford University Press, 2009). Important anthropological/microbial work continues to be done on antimicrobial resistance, for example, Clare Chandler's Anthropology of Microbial Resistance Research Group (https://www.lshtm.ac.uk/research/centres-projects-groups/anthropology-antimicrobial-resistance) at the London School of Hygiene and Tropical Medicine. Kathryn Orzech and Mark Nichter, "From resilience to resistance: Political ecological lessons from antibiotic and pesticide resistance," *Annual Review of Anthropology* 37 (2008): 267–82. As Hannah Landecker says, "antibiotic resistance is a collective ecological condition of late industrialism." Landecker, "Antibiotic Resistance and the Biology of History," 20.

38. Paxson and Helmreich, "Perils and Promises," 165.
39. Lynn Margulis and Dorion Sagan, *Microcosmos: Four Billion Years of Evolution from Our Microbial Ancestors* (Berkeley: University of California Press, 1997).
40. Donna Haraway, *Staying with the Trouble: Making Kin in the Chthulucene* (Durham, N.C.: Duke University Press, 2016), 60.
41. Donna Haraway, *When Species Meet* (Minneapolis: University of Minnesota Press, 2008).
42. Haraway, *Staying with the Trouble*, 4.
43. Ingold and Palsson, *Biosocial Becomings*, 39.
44. Ingold and Palsson, 39.
45. Maureen A. O'Malley and John Dupré, "Size Doesn't Matter: Towards a More Inclusive Philosophy of Biology," *Biology and Philosophy* 22 (2007): 155–91.
46. Nading, "Evidentiary Symbiosis," 562.
47. John Hartigan, *Aesop's Anthropology: A Multispecies Approach* (Minneapolis: University of Minnesota Press, 2014).
48. Jamie Lorimer, *The Probiotic Planet: Using Life to Manage Life* (Minneapolis: University of Minnesota Press, 2020).
49. Katherine R. Amato, Corinne F. Maurice, Karen Guillemin, and Tamara Giles-Vernick, "Multidisciplinarity in Microbiome Research: A Challenge and Opportunity to Rethink Causation, Variability, and Scale," *BioEssays* 41, no. 10 (2019): e1900007.
50. Suzanne L. Ishaq, Francisco J. Parada, Patricia G. Wolf, Carla Y. Bonilla, Megan A. Carney, Amber Benezra, Emily Wissel et al., "Introducing the Microbes and Social Equity Working Group: Considering the Microbial Components of Social, Environmental, and Health Justice," *mSystems* 6: e00471-21.
51. Christy A. Harrison and Douglas Taren, "How Poverty Affects Diet to Shape the Microbiota and Chronic Disease," *Nature Reviews Immunology* 18, no. 4 (2018): 279–87.

52. Beth Greenhough et al., "Setting the Agenda for Social Science Research on the Human Microbiome," *Palgrave Communication* 6, no. 18 (2020): Article 18.
53. https://sites.middlebury.edu/eggleston/.
54. The Amato Lab, https://www.kramato.com/.
55. Travis De Wolfe, Mohammed Rafi Arefin, Amber Benezra, and María Rebolleda Gómez, "Chasing Ghosts: Race, Racism, and the Future of Microbiome Research," *mSystems* 6: e00604-21.
56. William P. Hanage, "Microbiology: Microbiome Science Needs a Healthy Dose of Skepticism," *Nature* 512, no. 7514 (2014): 247–48.
57. J. D. Fortenberry, "The Uses of Race and Ethnicity in Human Microbiome Research," *Trends in Microbiology* 21, no. 4 (2013): 165–66. There is much more on race in chapter 4.
58. Keisha Findley, David R. Williams, Elizabeth A. Grice, and Vence L. Bonham, "Health Disparities and the Microbiome," *Trends in Microbiology* 24 (2016): 847–50.
59. Hannah Landecker, "Food as Exposure: Nutritional Epigenetics and the New Metabolism," *BioSocieties* 6 (2011): 167–94.
60. I thank Janelle Lamoreaux for writing "What If the Environment Is a Person? Lineages of Epigenetic Science in a Toxic China," *Cultural Anthropology* 31, no. 2 (2016): 188–214, and so many other things that made me think.

1. What We Talk About When We Talk About Collaboration

1. The Matsutake Worlds Research Group takes intra-action a step further by defining "intra-species," meant to convey a world in which things in relation do not preexist their relation: "A spore flies and lands on the soil where microbes dwell. A mycelium reaches out and entangles itself with plant roots. It collaborates with microbes and diversifies itself into different strands. Fungus and microbes dissolve the soil and create new strata and a new landscape. A mushroom as a fruit body blooms as a product of these intraspecies actions. . . . The ontologies of the things themselves are at stake in their relating." Matsutake Worlds Research Group, "A New Form of Collaboration in Cultural Anthropology: Matsutake Worlds," *American Ethnologist* 36, no. 2 (2009): 385.
2. Initially the project proposal ambitiously included me conducting ethnographic fieldwork at the Mal-ED site in Malawi as well.
3. My official, institutional partnership with the Gordon Lab lasted from 2010 to 2012, but coauthored papers were published into 2015.
4. Dr. Gordon insisted that I (and all the lab members) call him Jeff because of the intentionally open environment he fostered. Here it seems more appropriate to refer to him as Dr. Gordon.

5. Diana Forsythe operationalizing Laura Nader's term in *Studying Those Who Study Us: An Anthropologist in the World of Artificial Intelligence* (Stanford, Calif.: Stanford University Press, 2002).
6. Direct quote from a phone conversation with senior anthropology faculty.
7. Felicity Callard and Des Fitzgerald, *Rethinking Interdisciplinarity across the Social Sciences and Neurosciences* (London: Palgrave Macmillan, 2015).
8. Scott Frickel, Mathieu Albert, and Barbara Prainsack, eds., *Investigating Interdisciplinary Collaboration Theory and Practice across Disciplines* (New Brunswick, N.J.: Rutgers University Press, 2017).
9. Regina F. Bendix, Kilian Bizer, and Dorothy Noyes, *Sustaining Interdisciplinary Collaboration: A Guide for the Academy* (Chicago: University of Illinois Press, 2017).
10. Ana Viseu, "Integration of Social Science into Research Is Crucial," *Nature* 525, no. 291 (2015).
11. Alison Kraft and Beatrix Rubin, "Changing Cells: An Analysis of the Concept of Plasticity in the Context of Cellular Differentiation," *BioSocieties* 11, no. 4 (2016): 497–525; Hannah Landecker and Aaron Panofsky, "From Social Structure to Gene Regulation, and Back: A Critical Introduction to Environmental Epigenetics for Sociology," *Annual Review of Sociology* 39 (2013): 333–57; Jorg Niewöhner, "Epigenetics: Embedded Bodies and the Molecularisation of Biography and Milieu," *BioSocieties* 6, no. 3 (2011): 279–98.
12. Angela Willey, "Engendering New Materializations: Feminism, Nature, and the Challenge to Disciplinary Proper Objects," in *The New Politics of Materialism: History, Philosophy, Science*, edited by Sarah Ellenzweig and John H. Zammito (New York: Routledge, 2017), 316.
13. Sarah Franklin, *Biological Relatives: IVF, Stem Cells, and the Future of Kinship* (Durham, N.C.: Duke University Press, 2013); Rebecca M. Jordan-Young and Katrina Karkazis, *Testosterone: An Unauthorized Biography* (Cambridge, Mass.: Harvard University Press, 2020); Emily Martin, *Flexible Bodies: Tracking Immunity in American Culture from the Days of Polio to the Age of AIDS* (Boston: Beacon Press, 1994); Emilia Sanabria, *Plastic Bodies: Sex Hormones and Menstrual Suppression in Brazil* (Durham, N.C.: Duke University Press, 2016).
14. Meloni et al., *Palgrave Handbook of Biology and Society*, 2.
15. Diana Coole and Samantha Frost, *New Materialisms: Ontology, Agency, and Politics* (Durham, N.C.: Duke University Press, 2010); Rick Dolphijn and Iris van der Tuin, *New Materialism: Interviews and Cartographies* (Ann Arbor, Mich.: Open Humanities Press, 2012); Samantha Frost, "The Implications of the New Materialisms for Feminist Epistemology," in *Feminist Epistemology and Philosophy of Science*, edited by Heidi E. Grasswick, 69–83 (New York: Springer, 2011).

16. Banu Subramaniam and Angela Willey, "Introduction to Science out of Feminist Theory Part One: Feminism's Sciences," *Catalyst: Feminism, Theory, Technoscience* 3, no. 1 (2017): 12.
17. Banu Subramaniam, *Ghost Stories for Darwin: The Science of Variation and the Politics of Diversity* (Urbana, Ill.: University of Chicago Press, 2014), 5.
18. Banu Subramaniam and Angela Willey, "Introduction, Science out of Feminist Theory Part Two: Remaking Sciences," *Catalyst: Feminism, Theory, Technoscience* 3, no. 2 (2017): 8.
19. See Sara Ahmed, "Open Forum: Imaginary Prohibitions; Some Preliminary Remarks on the Founding Gestures of the 'New' Materialism," *European Journal of Women's Studies* 15, no. 1 (2008): 23–39, and response to Ahmed by Noela Davis, "New Materialism and Feminism's Antibiologism: A Response to Sara Ahmed," *European Journal of Women's Studies* 16, no. l (2009): 67–80.
20. Subramaniam, *Ghost Stories for Darwin,* 5.
21. Wilson, *Gut Feminism,* 175.
22. Wilson, 106.
23. Angela Willey, "Biopossibility: A Queer Feminist Materialist Science Studies Manifesto, with Special Reference to the Question of Monogamous Behavior," *Signs: Journal of Women in Culture and Society* 41, no. 3 (2016): 553–77.
24. Elizabeth F. S. Roberts, "Bio-ethnography: A Collaborative, Methodological Experiment in Mexico City," *Somatosphere,* February 25, 2015, http://somatosphere.net/2015/02/bio-ethnography.html; Roberts and Sanz, "Bioethnography."
25. Elizabeth F. S. Roberts, "When Nature/Culture Implodes: Feminist Anthropology and Biotechnology," in *Mapping Feminist Anthropology in the Twenty-First Century,* edited by E. Lewin and M. Silverstein, 105–25 (New Brunswick, N.J.: Rutgers University Press, 2016).
26. What Max Liboiron has referred to as "jurisdiction" (Discard Studies Conference, New York University, September 2022)—each of us within the power infrastructure of academia has our own jurisdiction, and although graduate students often would like to foment a revolution, revolution is outside our jurisdiction. More often than not, we have to do what our senior faculty and advisors tell us to do in order to get recommendations, postdocs, jobs, and so on.
27. More on metrics of malnutrition in chapter 3.
28. In the 1970s, Grameen Bank and its founder, Muhammad Yunus, started microfinance operations in Bangladesh—namely, "microcredit," which consist of small loans granted to poor women to start income-generating businesses out of their homes. Yunus and Grameen Bank won the Nobel Peace Prize in 2006 for microfinance, as it was seen as a social and

economic success (measured by the fact that their microcredit loans were repaid at a rate of 98 percent). However, anthropologist Lamia Karim's research shows that microcredit has been complicatedly disempowering for women, exposing a "contradiction between the rhetoric of microfinance organizations—that is, how these institutions produce 'truth' about their programs—and the lived realities of the women who were situated in dense webs of social and kin obligations and reciprocities that constrained their economic activities." Karim, *Microfinance and Its Discontents: Women in Debt in Bangladesh* (Minneapolis: University of Minnesota Press, 2011), 18.

29. Chapter 4 discusses the debate about potential placental microbe interactions, which backs first microbial interactions up into the uterus and has different implications.
30. See chapter 4 for more about mice.
31. For more about mice as experimental animals, see Rebecca Lemov, *World as Laboratory: Experiments with Mice, Mazes, and Men* (New York: Hill and Wang, 2005); Karen Rader, *Making Mice: Standardizing Animals for American Biomedical Research, 1900–1955* (Princeton, N.J.: Princeton University Press, 2004).
32. Nicole C. Nelson, *Model Behavior: Animal Experiments, Complexity, and the Genetics of Psychiatric Disorders* (Chicago: University of Chicago Press, 2018).
33. Nelson, 19.
34. Important Gordon Lab papers in the genealogy of this work that are foundational in the field of human microbial ecology include Lora V. Hooper, Dan R. Littman, and Andrew J. Macpherson, "Interactions between the Microbiota and the Immune System," *Science* 8 (2012): 1268–73; Ruth Ley, Peter Turnbaugh, Samuel Klein, and Jeffrey I. Gordon, "Human Gut Microbes Associated with Obesity," *Nature* 444, no. 7122 (2006): 1022–23; Pete Turnbaugh, Vanessa Ridaura, Jeremiah Faith, Federico Rey, Rob Knight, and Jeffrey I. Gordon, "The Effect of Diet on the Human Gut Microbiome: A Metagenomic Analysis in Humanized Gnotobiotic Mice," *Science Translational Medicine* 6, no. 14 (2009); Tanya Yatsunenko, Federico E. Rey, Mark J. Manary, Indi Trehan, Maria Gloria Dominguez-Bello, Monica Contreras, Magda Magris et al., "Human Gut Microbiome Viewed across Age and Geography," *Nature* 486, no. 7402 (2012): 222–27; Michelle I. Smith, Tanya Yatsunenko, Mark J. Manary, Indi Trehan, Rajhab Mkakosya, Jiye Cheng, Andrew L. Kau et al., "Gut Microbiomes of Malawian Twin Pairs Discordant for Kwashiorkor," *Science* 331 (2013): 548–54; Vanessa K. Ridaura, Jeremiah J. Faith, Federico E. Rey, Jiye Cheng, Alexis E. Duncan, Andrew L. Kau, Nicholas W. Griffin et al., "Gut Microbiota from Twins Discordant for Obesity Modulate Metabolism in Mice," *Science* 341 (2013): 1079–88; Sathish Subramanian, Sayeeda Huq, Tanya Yatsunenko, Rashidul Haque, Mustafa Mahfuz, Mohammed A. Alam, Amber Benezra et al., "Persistent

Gut Microbiota Immaturity in Malnourished Bangladeshi Children," *Nature* 510, no. 7505 (2014): 417–21.

35. This prize, as well as Dr. Gordon being named on the Reuters List in 2015 (this list comes from the intellectual property and science unit of Thomson Reuters, which predicts future Nobel laureates based on scientists' publication citations; the Reuters List has accurately forecasted 37 Nobel winners since 2002), and many other awards, seemed to point to an imminent Nobel nomination for Dr. Gordon. The attention the Gordon Lab received and the buzz around Dr. Gordon at this time put an unexpected strain on our collaboration.
36. More about this in chapter 3.
37. "Westernization" for the Gordon Lab is discussed more fully in chapter 3.
38. The number of lab members, as well as the number of lab members working on both malnutrition and Bangladesh has substantially increased in subsequent years.
39. Margaret Lock and Vinh-Kim Nguyen, *An Anthropology of Biomedicine* (Oxford: Wiley-Blackwell, 2010).
40. Koch, "Local Microbiologies of Tuberculosis."
41. João Biehl and Adriana Petryna, eds., *When People Come First: Critical Studies in Global Health* (Princeton, N.J.: Princeton University Press, 2013), 14.
42. All the icddr,b FRAS', mothers', and children's names throughout the text are pseudonyms. Their anonymity and privacy are very important to me, especially as they continue to be involved in icddr,b and Mal-ED studies. I randomly chose names from a list of the most popular Bangladeshi baby names of 2002.
43. Matsutake Worlds Research Group, "A New Form of Collaboration in Cultural Anthropology," 381.
44. Alan Hubbard, James Trostle, Ivan Cangemi, and Joseph N. S. Eisenberg, "Countering the Curse of Dimensionality: Exploring Data-Generating Mechanisms through Participant Observation and Mechanistic Modeling," *Epidemiology* 30, no. 4 (2019): 609–14; Thurka Sangaramoorthy, Amelia M. Jamison, and Typhanye V. Dyer, "HIV-Stigma, Retention in Care, and Adherence among Older Black Women Living with HIV," *Journal of Association of Nurses in AIDS Care* 28, no. 4 (2017): 518–31.
45. Mary Leighton and Elizabeth F. S. Roberts, "Trust/Distrust in Multidisciplinary Collaboration: Some Feminist Reflections," *Catalyst: Feminism, Theory, Technoscience* 6, no. 2 (2020): 1–27.
46. Forsythe, *Studying Those Who Study Us.*
47. Callard and Fitzgerald, *Rethinking Interdisciplinarity.*
48. Des Fitzgerald and Felicity Callard, "Entangled in the Collaborative Turn: Observations from the Field," *Somatosphere,* November 3, 2014, http://somatosphere.net/2014/11/entangled.html.
49. Yates-Doerr, "Whose Global, Which Health?"

50. Sarah Franklin, "Science as Culture, Cultures of Science," *Annual Review of Anthropology* 24 (1995): 163–84; Emily Martin, "Anthropology and the Cultural Study of Science," *Science, Technology, and Human Values* 23, no. 1 (1998): 24–44; Joan Fujimura, "Authorizing Science Studies and Anthropology," *American Anthropologist* 101, no. 2 (1999): 381–84; Michael M. J. Fischer, "Four Genealogies for a Recombinant Anthropology of Science and Technology," *Cultural Anthropology* 22, no. 4 (2007): 539–615.
51. Bruno Latour and Steve Woolgar, *Laboratory Life: The Construction of Scientific Facts* (Princeton, N.J.: Princeton University Press, 1986); Sharon Traweek, *Beamtimes and Lifetimes: The World of High Energy Physicists* (Cambridge, Mass.: Harvard University Press, 1988); Karin Knorr-Centina, *Epistemic Cultures: How the Sciences Make Knowledge* (Cambridge, Mass.: Harvard University Press, 1999); Park Doing, "Give Me a Laboratory and I Will Raise a Discipline: The Past, Present, and Future Politics of Laboratory Studies in STS," in *The Handbook of Science and Technology Studies*, 3rd ed., edited by Edward J. Hackett, Olga Amsterdamska, Michael E. Lynch, and Judy Wajcman, 279–96 (Cambridge, Mass.: MIT Press, 2007); Alison Cool, "Laboratories," in *Oxford Bibliographies in Anthropology*, edited by John L. Jackson (New York: Oxford University Press, 2016), https://doi.org/10.1093/OBO/9780199766567-0142.
52. Kim Fortun and Michael Fortun, "Scientific Imaginaries and Ethical Plateaus in Contemporary U.S. Toxicology," *American Anthropologist* 107, no. 1 (2005): 43–54.
53. Fortun and Fortun, 51.
54. Paul Rabinow and Gaymon Bennett, *Designing Human Practices: An Experiment with Synthetic Biology* (Chicago: University of Chicago Press, 2012).
55. Paul Rabinow, "Prosperity, Amelioration, Flourishing: From a Logic of Practical Judgment to Reconstruction," *Law and Literature* 21, no. 3 (2009): 301–20.
56. Rabinow, 304.
57. Leighton and Roberts, "Trust/Distrust in Multidisciplinary Collaboration," 5.
58. Callard and Fitzgerald, *Rethinking Interdisciplinarity*, 98–99.
59. All quotes are from conversations Dr. Gordon and I had—both formal ethnographic interviews and informal conversations.
60. Callard and Fitzgerald, *Rethinking Interdisciplinarity*, 105.
61. Helga Nowotny, Peter Scott, and Michael Gibbons, *Re-thinking Science: Knowledge and the Public in an Age of Uncertainty* (Cambridge: Polity Press, 2001).
62. Doris McGartland Rubio, Ellie E. Schoenbaum, Linda S. Lee, David E. Schteingart, Paul R. Marantz, Karl E. Anderson, Lauren Dewey Platt, Adriana Baez, and Karin Esposito, "Defining Translational Research:

Implications for Training," *Academic Medicine* 85, no. 3 (2016): 470–75.

63. *Journal of Translational Medicine*; *Science Translational Medicine*; Stephen H. Woolf, "The Meaning of Translational Research and Why It Matters," *Journal of the American Medical Association* 299, no. 2 (2008): 211–13.
64. Fortun and Fortun, "Scientific Imaginaries and Ethical Plateaus," 51.
65. Chris Kelty, "Ethnography and the IRB," *Savage Minds*, February 2006, https://savageminds.org/2006/02/08/ethnography-and-the-irb/; Rena Lederman, "Educate Your IRB (a Boilerplate Experiment)," *Savage Minds*, April 2007, http://savageminds.org/2007/04/02/educate-your-irb-a-boilerplate-experiment/; Kimberly Sue, "Are IRBs a Stumbling Block for an Engaged Anthropology?," *Somatosphere*, August 2012, http://somatosphere.net/2012/are-irbs-a-stumbling-block-for-an-engaged-anthropology.html/.
66. Fiona McDonald, Luke Cantarella, and Keith M. Murphy, "Responsibility," Theorizing the Contemporary, *Cultural Anthropology*, July 27, 2017, https://culanth.org/fieldsights/responsibility.
67. Ian Hacking, "The Complacent Disciplinarian," *Interdisciplines* (blog), July 27, 2008, http://ise-hec-cscs.blogspot.com/2009/07/interdisciplines-rethinking.html.
68. Marilyn Strathern, "Experiments in Interdisciplinarity," *Social Anthropology* 13, no. 1 (2005): 75–90; Nicolas Langlitz, "Out of the Armchair: On a Problem Shared by Anthropologists of Science and Philosophers of Mind," paper presented at the School for Advanced Research Fieldwork in Philosophy Conference, October 2012.
69. Funding acknowledgments for this paper include the NIH and the American Diabetes Association. Ashley R. Wolf, Darryl A. Wesener, Jiye Cheng, Alexandra N. Houston-Ludlam, Zachary W. Beller, Matthew C. Hibberd, Richard J. Giannone et al., "Bioremediation of a Common Product of Food Processing by a Human Gut Bacterium," *Cell Host Microbe* 26, no. 4 (2019): 463–77.
70. Wolf et al., 464.
71. Emily Yates-Doerr, "Translational Competency: On the Role of Culture in Obesity Interventions," *Medicine Anthropology Theory* 5, no. 4 (2018): 106–17.
72. It is not a coincidence that three of the four experiments to which I was most drawn in the lab were run by women and that my main interlocutors were women, even though the lab at that time was only about 15 percent female. In all fairness, the Gordon Lab was not consistently male dominated; the lab population tended to fluctuate over the years. Two women students of Dr. Gordon's, Ruth Ley and Lora Hooper, have gone on to be leaders in the field of human microbial ecology.
73. Nading, "Evidentiary Symbiosis," 561.
74. Nading, 560.

75. More about *pushti* packets in chapter 4.
76. Especially because I engage Indigenous thinking in chapter 3, it's important to note that none of my Bangladeshi interlocutors were Indigenous, nor did I meet any Adivasi people in Dhaka.
77. http://www.impatientoptimists.org/.
78. Carrie Friese, "Realizing Potential in Translational Medicine: The Uncanny Emergence of Care as Science," *Current Anthropology* 54, suppl. 7 (2013): S129–38.
79. See http://www.gatesfoundation.org/What-We-Do/Global-Health/Discovery-and-Translational-Sciences.
80. http://www.gatesfoundation.org/What-We-Do/Global-Health/Discovery-and-Translational-Sciences.
81. Yates-Doerr, "Translational Competency."
82. Yates-Doerr, 109.
83. http://www.icddrb.org/who-we-are/our-mission/about-us.
84. Melinda K. Munos, Christa L. Fischer Walker, and Robert E. Black, "The Effect of Oral Rehydration Solution and Recommended Home Fluids on Diarrhoea Mortality," *International Journal of Epidemiology* 39, suppl. 1 (2010): i75–87.
85. Michelle Murphy, *The Economization of Life* (Durham, N.C.: Duke University Press, 2017).
86. "Water with Sugar and Salt," *The Lancet* 312, no. 8084 (1978): 300–301.
87. Paul Farmer, "An Anthropology of Structural Violence," *Current Anthropology* 45, no. 3 (2004): 305–25.
88. Paul Farmer, Bruce Nizeye, Sara Stulac, and Salmaan Keshavjee, "Structural Violence and Clinical Medicine," *PLoS Medicine* 3, no. 10 (2006): 1689.
89. Farmer et al., 1687.
90. Emily Yates-Doerr, "Reworking the Social Determinants of Health: Responding to Material-Semiotic Indeterminacy in Public Health Interventions," *Medical Anthropology Quarterly* 34, no. 3 (2020): 380.
91. Tahmeed Ahmed, Rashidul Haque, Abul Mansur Shamsir Ahmed, William A. Petri Jr., and Alejandro Cravioto, "Use of Metagenomics to Understand the Genetic Basis of Malnutrition," *Nutrition Reviews* 67 (2009): 201–6.
92. Cal Biruk, "Ethical Gifts? An Analysis of Soap-for-Data Transactions in Malawian Survey Research Worlds," *Medical Anthropology Quarterly* 31, no. 3 (2017): 379.
93. Michel Callon, "Some Elements of a Sociology of Translation: Domestication of the Scallops and the Fishermen of St Brieuc Bay," *Sociological Review* 32, no. 1 (1984): 196–233.
94. Landecker, "Food as Exposure"; Nancy Scheper-Hughes, *Death without Weeping: The Violence of Everyday Life in Brazil* (Berkeley: University of California Press, 1992); Emily Yates-Doerr, *The Weight of Obesity* (Oakland:

University of California Press, 2015).

95. See chapters 3 and 4 for more about mothers.
96. Biruk, "Ethical Gifts?"
97. Benezra et al., "Anthropology of Microbes."
98. Bruno Latour, "Drawing Things Together," in *Representation in Scientific Practice,* edited by Michael Lynch and Steve Woolgar, 19–68 (Cambridge, Mass.: MIT Press, 1988).
99. Benezra et al., "Anthropology of Microbes," 6380.

2. How to Make a Microbiome

1. Annemarie Mol, *The Body Multiple* (Durham, N.C.: Duke University Press, 2002).
2. Toni Morrison, "Unspeakable Things Unspoken: The Afro-American Presence in American Literature," *Michigan Quarterly Review* 28, no. 1 (1989): 1–34.
3. Liboiron, *Pollution Is Colonialism,* 1.
4. Sara Ahmed, "Making Feminist Points," *feministkilljoys* (blog), September 11, 2013, https://feministkilljoys.com/2013/09/11/making-feminist-points/.
5. Sara Ahmed, *Living a Feminist Life* (Durham, N.C.: Duke University Press, 2017), 148.
6. Eve Tuck, K. Wayne Yang, and Rubén Gaztambide-Fernández, "Citation Practices," *Critical Ethnic Studies,* April 2015, http://www.criticalethnicstudiesjournal.org/citation-practices.
7. Max Liboiron, "Firsting in Research," *Discard Studies,* January 18, 2021, https://discardstudies.com/2021/01/18/firsting-in-research/.
8. John Ingraham, *March of the Microbes* (Boston: Harvard University Press, 2010).
9. Nancy Merino, Heidi S. Aronson, Diana P. Bojanova, Jayme Feyhl-Buska, Michael L. Wong, Shu Zhang, and Donato Giovannelli, "Living at the Extremes: Extremophiles and the Limits of Life in a Planetary Context," *Frontiers in Microbiology* 10 (April 15, 2019): 780.
10. I will try to keep calling back to fact that *firsting* in research is colonizing—"discovery" is a valued and lauded practice, but it erases existing knowledge and doesn't really exist. Liboiron, "Firsting in Research."
11. Maureen A. O'Malley and John Dupré, "Size Doesn't Matter: Towards a More Inclusive Philosophy of Biology," *Biology and Philosophy* 22 (2007): 159.
12. W. C. Summers, "From Culture as Organism to Organism as Cell: Historical Origins of Bacterial Genetics," *Journal of the History of Biology* 24, no. 2 (1991): 171–90.
13. Boris Magasanik, "A Midcentury Watershed: The Transition from Microbial Biochemistry to Molecular Biology," *Journal of Bacteriology* 181, no. 2

(1999): 357–58.

14. O'Malley and Dupré, "Size Doesn't Matter," 160.
15. O'Malley and Dupré, 161.
16. John Dupré, *Processes of Life: Essays in the Philosophy of Biology* (Oxford: Oxford University Press, 2012), 169.
17. O'Malley and Dupré, "Size Doesn't Matter," 162.
18. W. Ford Doolittle, "Uprooting the Tree of Life," *Scientific American,* February 2000, https://www.scientificamerican.com/article/uprooting-the-tree-of-life/.
19. Melissa B. Miller and Bonnie L. Bassler, "Quorum Sensing in Bacteria," *Annual Review of Microbiology* 55, no. 1 (2000): 165–99.
20. Joshua Lederberg and Alexa McCray, "'Ome Sweet 'Omics—a Genealogical Treasury of Words," *Scientist* 15, no. 7 (2001): 8.
21. C. M. Xochitl and Curtis Huttenhower, "Human Microbiome Analysis," *PLOS Computational Biology* 8, no. 12 (2012): 1–14.
22. Margaret McFall-Ngai, Michael G. Hadfield, Thomas C. G. Bosch, Hannah V. Carey, Tomislav Domazet-Lošo, Angela E. Douglas, Nicole Dubilier et al., "Animals in a Bacterial World: A New Imperative for the Life Sciences," *Proceedings of the National Academy of Sciences of the United States of America* 110, no. 9 (2013): 3229–36.
23. Astrid Schrader, "Responding to *Pfiesteria piscicida* (the Fish Killer): Phantomatic Ontologies, Indeterminacy, and Responsibility in Toxic Microbiology," *Social Studies of Science* 40, no. 2 (2010): 275–306.
24. Jeffrey I. Gordon, "Honor Thy Gut Symbionts Redux," *Science* 336, no. 6086 (2012): 1251–53.
25. Jonathan A. Eisen, "And Today in #microbiomania (aka Overselling the Microbiome)—Ridiculous Claim from Raphael Kellman's Book Marketing Group," *The Tree of Life* (blog), October 31, 2018, http://phylogenomics.blogspot.com/search/label/overselling%20the%20microbiome.
26. NIH National Human Genome Research Institute, "Talking Glossary of Genomic and Genetic Terms," http://www.genome.gov/glossary/.
27. Handelsman, "Metagenomics and Microbial Communities."
28. John Dupré and Maureen O'Malley, "Metagenomics and Biological Ontology," *Studies in History and Philosophy of Biological and Biomedical Sciences* 38, no. 4 (2007): 834.
29. Elizabeth K. Costello, Christian L. Lauber, Micah Hamady, Noah Fierer, Jeffrey I. Gordon, and Rob Knight, "Bacterial Community Variation in Human Body Habitats across Space and Time," *Science* 326, no. 5960 (2009): 1694–97.
30. Shaneen Leishman, Hong Lien Do, and Pauline J. Ford, "Cardiovascular Disease and the Role of Oral Bacteria," *Journal of Oral Microbiology* 2, no. 1

(2010).

31. Franco Frati, Cristina Salvatori, Cristoforo Incorvaia, Alessandro Bellucci, Giuseppe Di Cara, Francesco Marcucci, and Susanna Esposito, "The Role of the Microbiome in Asthma: The Gut–Lung Axis," *International Journal of Molecular Sciences* 20, no. 1 (2018): Article 123.
32. Megan Clapp, Nadia Aurora, Lindsey Herrera, Manisha Bhatia, Emily Wilen, and Sarah Wakefield, "Gut Microbiota's Effect on Mental Health: The Gut–Brain Axis," *Clinics and Practice* 7, no. 4 (2017): Article 987.
33. Bruno Bonaz, Thomas Bazin, and Sonia Pellissier, "The Vagus Nerve at the Interface of the Microbiota–Gut–Brain Axis," *Frontiers in Neuroscience* 12 (2018): Article 49.
34. Respectively, Beibei Yang, Jinbao Wei, Peijun Ju, and Jinghong Chen, "Effects of Regulating Intestinal Microbiota on Anxiety Symptoms: A Systematic Review," *General Psychiatry* 32, no. 2 (2019): e100056; Mireia Valles-Colomer, Gwen Falony, Youssef Darzi, Ettje F. Tigchelaar, Jun Wang, Raul Y. Tito, and Carmen Schiweck, "The Neuroactive Potential of the Human Gut Microbiota in Quality of Life and Depression," *Nature Microbiology* 4, no. 4 (2019): 623–32; Jennifer Fouquier, Nancy Moreno Huizar, Jody Donnelly, Cody Glickman, Dae-Wook Kang, Juan Maldonado, Rachel A. Jones et al., "The Gut Microbiome in Autism: Study-Site Effects and Longitudinal Analysis of Behavior Change," *mSystems* 6, no. 2 (2021): e00848-20.
35. Hooper et al., "Interactions between the Microbiota and the Immune System."
36. Martin Blaser, *Missing Microbes: How the Overuse of Antibiotics Is Fueling Our Modern Plagues* (New York: Picador Press, 2015).
37. Henry Haiser, David B. Gootenberg, Kelly Chatman, Gopal Sirasani, Emily P. Balskus, and Peter J. Turnbaugh, "Predicting and Manipulating Cardiac Drug Inactivation by the Human Gut Bacterium *Eggerthella Lenta*," *Science* 341, no. 6143 (2013): 295–98.
38. Noel Mueller, Maria Gloria Dominguez-Bello, Lawrence Appel, and Suchitra Hourigan, "'Vaginal Seeding' after a Caesarean Section Provides Benefits to Newborn Children," *BJOG: An International Journal of Obstetrics and Gynaecology* 127, no. 2 (2019): 301.
39. Tejeshwar Jain, Prateek Sharma, Abhi C. Are, Selwyn M. Vickers, and Vikas Dudeja, "New Insights into the Cancer–Microbiome–Immune Axis: Decrypting a Decade of Discoveries," *Frontiers in Immunology* 23 (February 2021).
40. https://evelobio.com/.
41. https://hmpdacc.org/ihmp/.
42. See chapters 3 and 4.

43. O'Malley and Dupré, "Size Doesn't Matter," 172.
44. McFall-Ngai et al., "Animals in a Bacterial World," 1.
45. Haraway, *When Species Meet*; Haraway, *Staying with the Trouble*; Stefan Helmreich, *Alien Ocean: Anthropological Voyages in Microbial Seas* (Berkeley: University of California Press, 2009); Helmreich, *Sounding the Limits of Life*; Hird, *Origins of Sociable Life*; Maureen O'Malley, *Philosophy of Microbiology* (Cambridge: Cambridge University Press, 2014); Paxson, "Post-Pasteurian Cultures"; Paxson, *The Life of Cheese* (Berkeley: University of California Press, 2013).
46. S. Eben Kirksey and Stefan Helmreich, "The Emergence of Multispecies Ethnography," *Cultural Anthropology* 25, no. 4 (2010): 545–76.
47. Haraway, *When Species Meet*, 17.
48. Haraway, 19.
49. Lynn Margulis and Dorion Sagan, *Dazzle Gradually: Reflections on the Nature of Nature* (White River Junction, Vt.: Chelsea Green, 2007), 28.
50. Margulis's *endosymbionic theory* proves that mitochondria and plastids (now organelles of eukaryotic cells) originated as separate prokaryotic organisms that were taken inside the cell as endosymbionts.
51. Margulis and Sagan, *Microcosmos*.
52. Paxson, "Post-Pasteurian Cultures," 15.
53. Haraway, *When Species Meet*, 30.
54. Sidney W. Mintz and Christine M. DuBois, "The Anthropology of Food and Eating," *Annual Review of Anthropology* 31 (2002): 99–119.
55. Haraway, *When Species Meet*, 165.
56. Karen Barad, *Meeting the Universe Halfway: Quantum Physics and the Entanglement of Matter and Meaning* (Durham, N.C.: Duke University Press, 2007).
57. There is an analysis to be made here about actor-network theory, symmetrical thinking in STS, and nonhumans, but I'm not sure I want to make it. Does anyone care about actor-network theory anymore? I'm grappling with the very (male) gendered, (white) raced legacy of actor-network theory specifically and STS generally. My dissertation was overwrought with citations of Latour, Woolgar, and Lynch, but I'm not interested in continuing to build with those bricks.
58. Karen Barad, "Agential Realism," in *The Science Studies Reader*, edited by Mario Biagioli, 1–11 (New York: Routledge, 1999); Bruno Latour, *Reassembling the Social: An Introduction to Actor-Network Theory* (Oxford: Oxford University Press, 2005); Mol, *Body Multiple*; Haraway, *When Species Meet*.
59. Dwayne Donald, "From What Does Ethical Relationality Flow? An 'Indian' Act in Three Artifacts," in *The Ecological Heart of Teaching: Radical Tales of Refuge and Renewal for Classrooms and Communities*, edited by J. Seidel and D. W. Jardine, 10–16 (New York: Peter Lang, 2016); Nicholas J. Reo, "Inawendiwin and Relational Accountability in Anishnaabeg Studies," *Journal of Ethnobiology* 39, no. 1 (2019): 65–75; Enrique Salmón, "Kincentric

Ecology: Indigenous Perceptions of the Human–Nature Relationship," *Ecological Applications* 10, no. 5 (2000): 1327–32; Vanessa Watts, "Indigenous Place-Thought and Agency amongst Humans and Non-humans (First Woman and Sky Woman Go on a European World Tour!)," *Decolonization: Indigeneity, Education, and Society* 2, no. 1 (2013): 20–34.

3. Microbiokinships

1. See, respectively, Haraway, *Staying with the Trouble*; Vanessa Agard-Jones, *Cultures of Energy* (podcast), episode 35, September 29, 2016, https://cenhs.libsyn.com/2016/09; Janelle Lamoreaux, "Toxicology and the Chemistry of Cohort Kinship," *Somatosphere*, January 17, 2020, http://somatosphere.net/2020/chemical-kinship.html/; Zoe Todd and Anja Kanngieser, "Attending to Environment as Kin Studies," *Constellations: Indigenous Contemporary Art from the Americas*, 2020, https://muac.unam.mx/constelaciones/; Katharine Dow and Janelle Lamoreaux, "Situated Kinmaking and the Population 'Problem,'" *Environmental Humanities* 12, no. 2 (2020): 475–91; Reena Shadaan and Michelle Murphy, "Endocrine-Disrupting Chemicals (EDCs) as Industrial and Settler Colonial Structures: Towards a Decolonial Feminist Approach," *Catalyst: Feminism, Theory, Technoscience* 6, no. 1 (2020): 1–36; Amy Moran-Thomas, "What Is Communicable? Unaccounted Injuries and 'Catching' Diabetes in an Illegible Epidemic," *Cultural Anthropology* 34 (2019): 471–502; Salmón, "Kincentric Ecology"; Watts, "Indigenous Place-Thought"; Donald, "From What Does Ethical Relationality Flow?"; Reo, "Inawendiwin and Relational Accountability in Anishnaabeg Studies."
2. Berg et al., "Microbiome Definition Re-visited."
3. TallBear, "Why Interspecies Thinking Needs Indigenous Standpoints."
4. Todd and Kanngieser, "Attending to Environment as Kin Studies."
5. Amber Benezra, "Race in the Microbiome," *Science, Technology, and Human Values* 45, no. 5 (2020): 877–902; De Wolfe et al., "Chasing Ghosts."
6. Janelle Baker, Paulla Ebron, Rosa Ficek, Karen Ho, Renya Ramirez, Zoe Todd, Anna Tsing, and Sarah E. Vaughn, "The Snarled Lines of Justice," *Orion Magazine*, Winter 2020, https://orionmagazine.org/article/the-snarled-lines-of-justice/.
7. Vanessa Agard-Jones, "Spray," *Somatosphere*, May 27, 2014, http://somatosphere.net/2014/spray.html/; Elizabeth F. S. Roberts, "Exposure," Theorizing the Contemporary, *Fieldsights*, June 28, 2017, https://culanth.org/fieldsights/exposure; Michelle Murphy, "Alterlife and Decolonial Chemical Relations," *Cultural Anthropology* 32, no. 4 (2017): 494–503; Rachel Lee, "A Lattice of Chemicalized Kinship: Toxicant Reckoning in a Depressive-Reparative Mode," *Catalyst: Feminism, Theory, Technoscience* 5, no. 2 (2020): 1–27.
8. Yates-Doerr, "Reworking the Social Determinants of Health."

9. Baker et al., "Snarled Lines of Justice."
10. The U.S. Centers for Disease Control and Prevention One Health office promotes an animal–human–ecosystem approach by "recognizing the interconnection between people, animals, plants, and their shared environment" (https://www.cdc.gov/onehealth/index.html). Joe Copper Jack, Jared Gonet, Anne Mease, and Katarzyna Nowak, "Traditional Knowledge Underlies One Health," *Science* 369, no. 6511 (2020): 1576.
11. If I am doing this all wrong, if I am causing harm, I would like to know.
12. Omar Delannoy-Bruno, Chandani Desai, Arjun S. Raman, Robert Y. Chen, Matthew C. Hibberd, Jiye Cheng, Nathan Han et al., "Evaluating Microbiome-Directed Fibre Snacks in Gnotobiotic Mice and Humans," *Nature* 595 (2021): 91–95.
13. Michelle Murphy, "What Can't a Body Do?," *Catalyst: Feminism, Theory, Technoscience* 3, no. 1 (2017): 1–15.
14. As discussed further in chapter 4, "small" is determined by the Z-score classification metric laid out in the WHO Global Database on Child Growth and Malnutrition. Z-score interprets weight-for-height, height-for-age, and weight-for-age. Since the 1990s, Z-score classification has been challenged as an inaccurate locally or geographically relative indicator of child growth.
15. Dow and Lamoreaux, "Situated Kinmaking," 483.
16. V. K. Nguyen, "Government-by-Exception: Enrolment and Experimentality in Mass HIV Treatment Programmes in Africa," *Social Theory and Health* 7 (2009): 197.
17. Kasia Paprocki, *Threatening Dystopias: The Global Politics of Climate Change Adaptation in Bangladesh* (Ithaca, N.Y.: Cornell University Press, 2021).
18. Murphy, *Economization of Life,* 104.
19. https://www.icddrb.org/dmdocuments/icddr,b%20strategic%20plan%20 2019-2022_16June19.pdf.
20. Elizabeth F. S. Roberts, "Bioethnography and the Birth Cohort: A Method for Making New Kinds of Anthropological Knowledge about Transmission (Which Is What Anthropology Has Been About All Along)," *Somatosphere,* November 19, 2019, http://somatosphere.net/2019/bio ethnography-anthropological-knowledge-transmission.html/.
21. Laura Senier, Phil Brown, Sara Shostak, and Bridget Hanna, "The Socio-exposome: Advancing Exposure Science and Environmental Justice in a Post-genomic Era," *Environmental Sociology* 3, no. 2 (2017): 107.
22. United Nations Human Settlements Programme, "The Slums of the World: the face of urban poverty in the new millennium?," UN-HABITAT working paper, 2003, https://unhabitat.org/slums-of-the-world-the-face -of-urban-poverty-in-the-new-millennium.
23. Liboiron, *Pollution Is Colonialism.*
24. Shadaan and Murphy, "Endocrine-Disrupting Chemicals."
25. Moran-Thomas, "What Is Communicable?," 489.

26. Md. Khalid Hasan, Abrar Shahriar, and Kudrat Ullah Jim, "Water Pollution in Bangladesh and Its Impact on Public Health," *Heliyon* 5, no. 8 (2019): e02145.
27. While contaminants, toxins, and pathogens in drinking water are seemingly obvious and important for gut microbiome research, I could find only one article that addresses this issue in humans: Ruth Bowyer, Daniel N. Schillereff, Matthew A. Jackson, Caroline Le Roy, Philippa M. Wells, Tim D. Spector, and Claire J. Steves, "Associations between UK Tap Water and Gut Microbiota Composition Suggest the Gut Microbiome as a Potential Mediator of Health Differences Linked to Water Quality," *Science of the Total Environment* 739 (October 2020).
28. Yates-Doerr, "Reworking the Social Determinants of Health," 380.
29. Mol, *Body Multiple.*
30. Landecker, "Antibiotic Resistance and the Biology of History," 20.
31. Harris Solomon, *Metabolic Living: Food, Fat, and the Absorption of Illness in India* (Durham, N.C.: Duke University Press, 2016).
32. Solomon, 9.
33. Pierre DeMeyts and Nathalie Delzenne, "The Brain–Gut–Microbiome Network in Metabolic Regulation and Dysregulation," *Frontiers of Endocrinology* 12 (September 2021); Richard W. Stephens, Lidia Arhire, and Mihai Covasa, "Gut Microbiota: From Microorganisms to Metabolic Organ Influencing Obesity," *Obesity* 26, no. 5 (2018): 801–9.
34. Marilyn Strathern, *Partial Connections,* Updated ed. (Walnut Creek, Calif.: AltaMira Press, 2004), xiv.
35. Strathern, xxix.
36. Mol, *Body Multiple.*
37. Yatsunenko et al., "Human Gut Microbiome," 224.
38. See note 7.
39. Landecker, "Food as Exposure."
40. Simpson, "Not Murdered, Not Missing."
41. Landecker, "Antibiotic Resistance and the Biology of History," 19.
42. Braden T. Tierney, Zhen Yang, Jacob M. Luber, Marc Beaudin, Marsha C. Wibowo, Christina Baek, Eleanor Mehlenbacher, Chirag J. Patel, and Aleksandar D. Kostic, "The Landscape of Genetic Content in the Gut and Oral Human Microbiome," *Cell, Host, and Microbe* 26, no. 2 (2019): 283–95.
43. Yatsunenko et al., "Human Gut Microbiome," 227.
44. There are also the fields of environmental microbiomics, microbiome of the built environment, and host–microbe interactions, which all deal in various "environments."
45. Adele Clarke and Donna Haraway, eds., *Making Kin Not Population: Reconceiving Generations* (Chicago: Prickly Paradigm Press, 2018).
46. Banu Subramaniam, "'Overpopulation' Is Not the Problem," November 27, 2018, https://www.publicbooks.org/overpopulation-is-not-the-problem/.
47. Dow and Lamoreaux, "Situated Kinmaking."

48. Ruha Benjamin, "Black AfterLives Matter," in Clarke and Haraway, *Making Kin*, 42.
49. Benjamin, 42.
50. Kim TallBear, "Making Love and Relations beyond Settler Sex and Family," in Clarke and Haraway, *Making Kin*, 147.
51. Lamoreaux, "Toxicology and the Chemistry of Cohort Kinship."
52. Lauren Beck, ed., *Firsting in the Early Modern-Atlantic World* (New York: Routledge, 2020).
53. Liboiron, "Firsting in Research."
54. It will be immediately apparent to some that the Indigenous scholars discussed here are all from what are now called North and South America, and in this way, I am reproducing the same kinds of erasures of Indigenous and Aboriginal voices from all the other parts of the globe. I also realize that very few nonscientific Bangladeshi authors are cited. I am sorry for this.
55. Donald, "From What Does Ethical Relationality Flow?," 11.
56. Salmón, "Kincentric Ecology," 1332.
57. Reo, "Inawendiwin and Relational Accountability in Anishnaabeg Studies," 65.
58. Jeremiah J. Faith, Jean-Frédéric Colombel, and Jeffrey I. Gordon, "Identifying Strains That Contribute to Complex Diseases through the Study of Microbial Inheritance," *Proceedings of the National Academy of Sciences of the United States of America* 112, no. 3 (2015): 635.
59. Laura Grieneisen, Mauna Dasari, Trevor J. Gould, Johannes R. Björk, Jean Christophe Grenier, Vania Yotova, David Jansen et al., "Gut Microbiome Heritability Is Near-Universal but Environmentally Contingent," *Science* 373, no. 6551 (2021): 181–86.
60. Franklin, *Biological Relatives*; Franklin, *Dolly Mixtures: The Remaking of Genealogy* (Durham, N.C.: Duke University Press, 2007); Franklin, "Re-thinking Nature–Culture: Anthropology and the New Genetics," *Anthropological Theory* 3, no. 1 (2003): 65–85.
61. Maria G. Dominguez-Bello, Elizabeth K. Costello, Monica Contreras, Magda Magris, Glida Hidalgo, Noah Fierer, and Rob Knight, "Delivery Mode Shapes the Acquisition and Structure of the Initial Microbiota across Multiple Body Habitats in Newborns," *Proceedings of the National Academy of Sciences of the United States of America* 107, no. 26 (2010): 11971–75.
62. Bruce German, Samara Freeman, Carlito Lebrilla, and David A. Mills, "Human Milk Oligosaccharides: Evolution, Structures and Bioselectivity as Substrates for Intestinal Bacteria," *Nestlé Nutrition Institute Workshop Series: Pediatric Program* 62 (2008): 205–22.
63. Haraway, *When Species Meet*.
64. Referring again to Nading's idea that the category of "microbiome" circulates only in the Global North (see chapter 1), this makes me wonder at this late date, does the microbiome exist in Mirpur?

65. Seuty Sabur, "The Limits of Radical Politics in an Unstable 'Field': Rethinking Shahabag, Hefazat-e-Islam, and the Women's Grand Rally," *Fieldsights*, March 16, 2021, https://culanth.org/fieldsights/the-limits-of-radical-politics-in-an-unstable-field-rethinking-shahabag-hefazat-e-islam-and-the-womens-grand-rally.
66. Lamia Karim, "Analyzing Women's Empowerment: Microfinance and Garment Labor in Bangladesh," *Fletcher Forum of World Affairs* 38, no. 2 (2014): 153–66.
67. Yates-Doerr, *Weight of Obesity*, 169.
68. Watts, "Indigenous Place-Thought."
69. Styres, "Literacies of Land."
70. Watts, "Indigenous Place-Thought," 21.
71. Shadaan and Murphy, "Endocrine-Disrupting Chemicals," 8.
72. TallBear, "Why Interspecies Thinking Needs Indigenous Standpoints," 5.
73. More about malnutrition metrics in chapter 4.
74. Koch, "Local Microbiologies of Tuberculosis."
75. See chapter 4 and Amber Benezra, "Datafying Microbes: Malnutrition at the Intersection of Genomics and Global Health," *BioSocieties* 11 (2016): 334–51.
76. Rapp, *Testing Women, Testing the Fetus*.
77. Kjersti Aagaard, Jun Ma, Kathleen M. Antony, Radhika Ganu, Joseph Petrosino, and James Versalovic, "The Placenta Harbors a Unique Microbiome," *Science Translational Medicine* 6, no. 237 (2014): 237–65; Lisa Funkhouser and Seth Bordenstein, "Mom Knows Best: The Universality of Maternal Microbial Transmission," *PLoS Biology* 11 (2013): 1–9; Lisa F. Stinson, Mary C. Boyce, Matthew S. Payne, and Jeffrey A. Keelan, "The Not-So-Sterile Womb: Evidence That the Human Fetus Is Exposed to Bacteria prior to Birth," *Frontiers in Microbiology*, June 4, 2019.
78. Marcus C. de Goffau, Susanne Lager, Ulla Sovio, Francesca Gaccioli, Emma Cook, Sharon J. Peacock, Julian Parkhill, D. Stephen Charnock-Jones, and Gordon C. S. Smith, "Human Placenta Has No Microbiome but Can Contain Potential Pathogens," *Nature* 572, no. 7769 (2019): 329–34; Abigail P. Lauder, Aoife M. Roche, Scott Sherrill-Mix, Aubrey Bailey, Alice L. Laughlin, Kyle Bittinger, Rita Leite, Michal A. Elovitz, Samuel Parry, and Frederic D. Bushman, "Comparison of Placenta Samples with Contamination Controls Does Not Provide Evidence for a Distinct Placenta Microbiota," *Microbiome* 4, no. 1 (2016): Article 29.
79. Ed Yong, "Why the Placental Microbiome Should Be a Cautionary Tale," *The Atlantic*, July 31, 2017, https://www.theatlantic.com/science/archive/2019/07/placental-microbiome-should-be-cautionary-tale/595114/.
80. Becky Mansfield and Julie Guthman, "Epigenetic Life: Biological Plasticity, Abnormality, and New Configurations of Race and Reproduction," *Cultural Geographies* 22, no. 1 (2015): 3–20.

81. Mansfield and Guthman, 15.
82. Martine Lappé and Robbin Jeffries Hein, "Human Placenta, Birth Cohorts, and the Production of Epigenetic Knowledge," *Somatosphere,* February 27, 2020, http://somatosphere.net/2020/placenta-epigenetics-knowledge.html/.
83. Maria Gloria Dominguez-Bello, Filipa Godoy-Vitorino, Rob Knight, and Martin J. Blaser, "Role of the Microbiome in Human Development," *Gut* 68, no. 6 (2019): 1109, emphasis added.
84. Helmreich, *Alien Ocean,* 104.
85. Yates-Doerr, "Reworking the Social Determinants of Health."

4. Malnutrition Futures

1. Arturo Escobar, *Encountering Development: The Making and Unmaking of the Third World* (Princeton, N.J.: Princeton University Press, 1995).
2. Andrew Balmer, Jane Calvert, Claire Marris, Susan Molyneux-Hodgson, Emma Frow, Matthew Kearnes, Kate Bulpin, Pablo Schyfter, Adrian MacKenzie, and Paul Martin, "Taking Roles in Interdisciplinary Collaborations: Reflections on Working in Post-ELSI Spaces in the UK Synthetic Biology Community," *Science and Technology Studies* 28, no. 3 (2015): 3–25.
3. Fortun and Fortun, "Scientific Imaginaries and Ethical Plateaus."
4. Viktor Mayer-Schoenberger and Kenneth Cukier, *Big Data: A Revolution That Will Transform How We Live, Work, and Think* (London: John Murray, 2013).
5. Jose van Dijck, "Datafication, Dataism, and Dataveillance: Big Data between Scientific Paradigm and Ideology," *Surveillance and Society* 12, no. 2 (2014): 197–208.
6. Geoffrey Bowker, "The Theory/Data Thing," *International Journal of Communication* 8, no. 2043 (2012): 1795–99; Lisa Gitelman, ed., *"Raw Data" Is an Oxymoron* (Cambridge, Mass.: MIT Press, 2012); Leonelli, "Global Data for Local Science"; Antonia Walford, "Data Moves: Taking Amazonian Climate Science Seriously," *Cambridge Anthropology* 30, no. 2 (2012): 101–17.
7. Bowker, "Theory/Data Thing," 1797.
8. Sabrina Leonelli and R. A. Ankeny, "Re-thinking Organisms: The Impact of Databases on Model Organism Biology," *Studies in History and Philosophy of Biological and Biomedical Sciences* 43, no. 1 (2012): 29–36.
9. Biehl and Petryna, *When People Come First.*
10. Callard and Fitzgerald, *Rethinking Interdisciplinarity.*
11. More about *pushti* packets in chapter 1.
12. http://www.plumpyfield.com/file/resources/plaquette-ppf-2017-en.pdf.
13. Carolyn Lesorogol, Sherlie Jean-Louis, Jamie Green, and Lora Ianotti, "Preventative Lipid-Based Nutrient Supplements (LNS) and Young Child Feeding Practices: Findings from Qualitative Research in Haiti," *Maternal*

and Child Nutrition 11, no. 4 (2015): 62–76.

14. Tahmeed Ahmed, Mustafa Mahfuz, Santhia Ireen, and A. M. Shamsir Ahmed, "Nutrition of Children and Women in Bangladesh: Trends and Directions for the Future," *Journal of Health, Population, and Nutrition* 1 (2012): 1–11.
15. Sally Craddock, *Retired Except on Demand: The Life of Dr. Cicely Williams* (Oxford: Oxford University Press, 1984).
16. Michael Worboys, "The Discovery of Colonial Malnutrition between the Wars," in *Imperial Medicine and Indigenous Societies,* edited by David Arnold (Manchester, U.K.: Manchester University Press, 1988), 208.
17. Vincanne Adams, ed., *Metrics: What Counts in Global Health* (Durham, N.C.: Duke University Press, 2016).
18. Landecker, "Antibiotic Resistance and the Biology of History," 37.
19. WHO Global Database on Child Growth and Malnutrition, https://www.who.int/teams/nutrition-and-food-safety/databases/nutgrowthdb.
20. Stephanie A. Richard, Benjamin J. J. McCormick, Mark A. Miller, Laura E. Caulfield, William Checkley, and MAL-ED Network Investigators, "Modeling Environmental Influences on Child Growth in the MAL-ED Cohort Study: Opportunities and Challenges," *Clinical Infectious Diseases* 59 (2014): S255–60.
21. E.g., E. F. S. Roberts, "Food Is Love: And So, What Then?," *BioSocieties* 10, no. 2 (2015): 247–52; Yates-Doerr, *Weight of Obesity*; and Solomon, *Metabolic Living.*
22. Rayna Rapp, "Chasing Science: Children's Brains, Scientific Inquiries, and Family Labors," *Science, Technology, and Human Values* 36, no. 5 (2011): 674.
23. Martine Lappé, "Taking Care: Anticipation, Extraction and the Politics of Temporality in Autism Science," *BioSocieties* 9, no. 3 (2014): 304–28.
24. Lock and Nguyen, *An Anthropology of Biomedicine,* 90.
25. Koch, "Local Microbiologies of Tuberculosis," 96.
26. More about environments in chapter 3.
27. Smith et al., "Gut Microbiomes of Malawian Twin Pairs."
28. Emilia Sanabria and Emily Yates-Doerr, "Alimentary Uncertainties: From Contested Evidence to Policy," *BioSocieties* 10, no. 2 (2015): 117–24.
29. Sanabria and Yates-Doerr, 120.
30. Wolfgang-Peter Zingel, Markus Keck, Benjamin Etzold, and Hans-Georg Bohle, "Urban Food Security and Health Status of the Poor in Dhaka, Bangladesh," in *Health in Megacities and Urban Areas,* edited by Alexander Krämer, Mobarak Hossain Khan, and Frauke Kraas, 301–19 (Heidelberg, Germany: Springer, 2011).
31. Ian Whitmarsh, "The Ascetic Subject of Compliance: The Turn to Chronic Diseases in Global Health," in Biehl and Petryna, *When People*

Come First, 302–24.

32. Jeffrey Gordon, Nancy Knowlton, David A. Relman, Forest Rohwer, and Merry Youle, "Superorganisms and Holobionts," *Microbe* 8, no. 4 (2013): 152–53.
33. Lynda Birke, "Animal Bodies in the Production of Scientific Knowledge: Modelling Medicine," *Body and Society* 18 (2012): 156–78; Haraway, *When Species Meet*; Robert Kohler, *Lords of the Fly: Drosophila Genetics and the Experimental Life* (Chicago: University of Chicago Press, 1993); Lemov, *World as Laboratory*; Rader, *Making Mice.*
34. Michael E. Lynch discusses the strange but significant laboratory use of the term *sacrifice* for killing animals and how the biological body of the experimental animal is transformed to an object with "transcendental" meaning for science-at-large. Lynch, "Sacrifice and the Transformation of the Animal Body into a Scientific Object: Laboratory Culture and Ritual Practice in the Neurosciences," *Social Studies of Science* 18, no. 2 (1988): 265–89.
35. Turnbaugh et al., "Effect of Diet on the Human Gut Microbiome."
36. Daniel Engber, "The Trouble with Black-6: Black-6 Lab Mice and the History of Biomedical Research," *Slate*, November 17, 2011, http://www.slate.com/articles/health_and_science/the_mouse_trap/2011/11/black_6_lab_mice_and_the_history_of_biomedical_research.html.
37. Rader, *Making Mice.*
38. Alison Cool, "Twins Nature and Nurture," *BioSocieties* 9, no. 3 (2014): 225–27.
39. Fecal transplants are the transfer of stool from a healthy donor into a person's gastrointestinal tract for the purpose of treating recurrent *C. difficile*, the bacteria commonly responsible for diarrheal disease following antibiotic treatment. *C. difficile* infection is increasingly a problem in inpatient health care facilities where patients have been treated with broad-spectrum antibiotics and all of their commensal microbes have been wiped out. Lawrence Brandt and Olga Aroniadis, "An Overview of Fecal Microbiota Transplantation: Techniques, Indications, and Outcomes," *Gastrointestinal Endoscopy* 78, no. 2 (2013): 240–49.
40. Nelson, *Model Behavior*, 13.
41. Illumina, "An Introduction to Next-Generation Sequencing Technology," https://www.illumina.com/content/dam/illumina-marketing/documents/systems/miseq/Introduction_to_Next-Generation_Sequencing_Technology.pdf.
42. Adam Hedgecoe, *Politics of Personalised Medicine: Pharmacogenetics in the Clinic* (Cambridge: Cambridge University Press, 2004).
43. Vincenne Adams, Michelle Murphy, and Adele Clarke, "Anticipation: Technoscience, Life, Affect, Temporality," *Subjectivity* 28 (2009): 246–65.

44. Hedgecoe, *Politics of Personalised Medicine*, 17–18.
45. Xochitl and Huttenhower, "Human Microbiome Analysis."
46. Adams et al., "Anticipation," 247.
47. Sarah Richardson and Hallam Stevens, eds., *Postgenomics: Perspectives on Biology after the Genome* (Durham, N.C.: Duke University Press, 2015).
48. Hanage, "Microbiology," 248.
49. Hannah Landecker, "Being and Eating: Losing Grip on the Equation," *BioSocieties* 10, no. 2 (2015): 253–58.
50. Landecker, 257.
51. Lawrence A. David, Corinne F. Maurice, Rachel N. Carmody, David B. Gootenberg, Julie E. Button, Benjamin E. Wolfe, Alisha V. Ling et al., "Diet Rapidly and Reproducibly Alters the Human Gut Microbiome," *Nature* 505, no. 7484 (2014): 559–63; Turnbaugh et al., "Effect of Diet on the Human Gut Microbiome."
52. Andrew L. Goodman, George Kallstrom, Jeremiah J. Faith, Alejandro Reyes, Aimee Moore, Gautam Dantas, Jeffrey I. Gordon et al., "Extensive Personal Human Gut Microbiota Culture Collections Characterized and Manipulated in Gnotobiotic Mice," *Proceedings of the National Academy of Sciences of the United States of America* 108, no. 15 (2011): 6252–57.
53. Ridaura et al., "Gut Microbiota from Twins Discordant for Obesity."
54. Ian Whitmarsh, "Troubling 'Environments': Postgenomics, Bajan Wheezing, and Levi-Strauss," *Medical Anthropology Quarterly* 27, no. 4 (2013): 489–509.
55. Landecker, "Food as Exposure."
56. Sara Shostak, "Locating Gene–Environment Interaction: At the Intersections of Genetics and Public Health," *Social Science and Medicine* 56 (2003): 2327–42; Shostak, *Exposed Science: Genes, the Environment, and the Politics of Population Health* (Berkeley: University of California Press, 2013).
57. Subramanian et al., "Persistent Gut Microbiota Immaturity."
58. Sheila Jasanoff, "Virtual, Visible, and Actionable: Data Assemblages and the Sightlines of Justice," *Big Data and Society* 4, no. 2 (2017): 1–15; Kelly Bronson and Irena Knezevic, "Big Data in Food and Agriculture," *Big Data and Society* 3, no. 1 (2016): 1–5; Daniel McDonald, Gustavo Glusman, and Nathan Price, "Personalized Nutrition through Big Data," *Nature Biotechnology* 34 (2016): 152–54.
59. More about mothers and responsibility in chapter 3.
60. Sathish Subramanian, Laura Blanton, Steven A. Frese, Mark Charbonneau, David A. Mills, and Jeffrey I. Gordon, "Cultivating Healthy Growth and Nutrition through the Gut Microbiota," *Cell* 161, no. 1 (2015): 39.
61. Subramanian et al., "Persistent Gut Microbiota Immaturity," 420.
62. Aleksandra Kolodziejczyk, Danping Zheng, and Eran Elinav, "Diet–Microbiota Interactions and Personalized Nutrition," *Nature Reviews Microbiology* 17, no. 12 (2019): 742–53.

63. Kolodziejczyk et al., 742.
64. Jeanette L. Gehrig, Siddarth Venkatesh, Hao-Wei Chang, Matthew C. Hibberd, Vanderlene L. Kung, Jiye Cheng, Robert Y. Chen et al., "Effects of Microbiota-Directed Foods in Gnotobiotic Animals and Undernourished Children," *Science* 365, no. 6649 (2019): eaau4732.
65. Ishita Mostafa, Naila Nurun Nahar, Md. Munirul Islam, Sayeeda Huq, Mahfuz Mustafa, Michael Barratt, Jeffrey I. Gordon, and Tahmeed Ahmed, "Proof-of-Concept Study of the Efficacy of a Microbiota-Directed Complementary Food Formulation (MDCF) for Treating Moderate Acute Malnutrition," *BMC Public Health* 20 (2020): Article 242.
66. Robert Y. Chen, Ishita Mostafa, Matthew C. Hibberd, Subhasish Das, Mustafa Mahfuz, Nurun N. Naila, M. Munirul Islam et al., "A Microbiota-Directed Food Intervention for Undernourished Children," *New England Journal of Medicine* 384, no. 16 (2021): 1517–28.
67. Robert Y. Chen, Ishita Mostafa, Matthew C. Hibberd, Subhasish Das, Hannah M. Lynn, Daniel M. Webber, Mustafa Mahfuz, Michael J. Barratt, Tahmeed Ahmed, and Jeffrey I. Gordon, "Melding Microbiome and Nutritional Science with Early Child Development," *Nature Medicine* 27, no. 9 (2021): 1500–1509.
68. Chen et al., 1509.
69. Landecker, "Food as Exposure."
70. Jonathan M. Green, Michael J. Barratt, Michael Kinch, and Jeffrey I. Gordon, "Food and Microbiota in the FDA Regulatory Framework," *Science* 357, no. 6346 (2017): 39–40; Mikko Jauho and Mari Niva, "Lay Understandings of Functional Foods as Hybrids of Food and Medicine," *Food, Culture, and Society* 16 (2013): 43–63; Kate Weiner and Catherine Will, "Materiality Matters: Blurred Boundaries and the Domestication of Functional Foods," *BioSocieties* 10, no. 2 (2015): 194–212.
71. Melody J. Slashinski, Sheryl A. McCurdy, Laura S. Achenbaum, Simon N. Whitney, and Amy L. McGuire, "'Snake-Oil,' 'Quack Medicine,' and 'Industrially Cultured Organisms': Biovalue and the Commercialization of Human Microbiome Research," *BioMed Central Medical Ethics* 13, no. 28 (2012): 6.
72. Catherine Waldby, *The Visible Human Project: Informatic Bodies and Posthuman Medicine* (New York: Routledge, 2000).
73. Green et al., "Food and Microbiota in the FDA Regulatory Framework," 39–40.
74. More about *B. infantis* in chapter 3.
75. https://www.evivo.com/.
76. Subramanian et al., "Cultivating Healthy Growth and Nutrition," 44.
77. Biehl and Petryna, *When People Come First*, 14.

5. Ghosting Race

1. Ann Morning, "'Everyone Knows It's a Social Construct': Contemporary Science and the Nature of Race," *Sociological Focus* 40, no. 44 (2007): 436–54.
2. Ruha Benjamin, *Race after Technology* (Cambridge: Polity Press, 2019); Alondra Nelson, *The Social Life of DNA: Race, Reparations, and Reconciliation after the Genome* (Boston: Beacon Press, 2016); Fullwiley; Morning; Roberts; and many others.
3. John Hartigan, ed., *Anthropology of Race: Genes, Biology, and Culture* (Santa Fe, N.M.: School for Advanced Research Press, 2013).
4. Faye Harrison, "Introduction: Expanding the Discourse on 'Race,'" *American Anthropologist* 100, no. 3 (1998): 609–31; Barbara A. Koening, Sandra Soo-Jin Lee, and Sarah Richardson, *Revisiting Race in a Genomic Age* (New Brunswick, N.J.: Rutgers University Press, 2008); Leith Mullings, "Interrogating Racism: Toward an Antiracist Anthropology," *Annual Review of Anthropology* 34 (2005): 667–93.
5. Ian Whitmarsh and David S. Jones, eds., *What's the Use of Race? Modern Governance and the Biology of Difference* (Cambridge, Mass.: MIT Press, 2010).
6. Haraway, *When Species Meet.*
7. Many thanks to Katrina and Beck for creating the scaffolding on which this chapter and other work of mine builds. Katrina Karkazis and Rebecca Jordan-Young, "Sensing Race as a Ghost Variable in Science Technology, and Medicine," *Science, Technology, and Human Values* 45, no. 5 (2020): 763.
8. Dorothy Roberts, *Fatal Invention: How Science, Politics, and Big Business Re-create Race in the Twenty-First Century* (New York: New Press, 2011).
9. Hartigan, *Anthropology of Race*, 194.
10. Carol Mukhopadhyay and Yolanda Moses, "Reestablishing 'Race' in Anthropological Discourse," *American Anthropologist* 99, no. 3 (1997): 517–33.
11. Fatimah Jackson, "Ethnogenetic Layering (EL): An Alternative to the Traditional Race Model in Human Variation and Health Disparity Studies," *Annals of Human Biology* 35, no. 2 (2008): 121–44.
12. Barad, *Meeting the Universe Halfway.*
13. Amade M'charek, Katharina Schramm, and David Skinner, "Introduction: Technologies of Belonging—The Absent Presence of Race in Europe," *Science, Technology, and Human Values* 39, no. 4 (2014): 459–67.
14. Gabriela K. Fragiadakis, Samuel A. Smits, Erica D. Sonnenburg, William Van Treuren, Gregor Reid, Rob Knight, Alphaxard Manjurano et al., "Links between Environment, Diet, and the Hunter-Gatherer Microbiome," *Gut Microbes* 10, no. 2 (2019): 216–27.
15. Jacques Ravel, Pawel Gajer, Zaid Abdo, G. Maria Schneider, Sara S. K. Koenig, Stacey L. McCulle, Shara Karlebach et al., "Vaginal Microbiome of Reproductive-Age Women," *Proceedings of the National Academy of Sciences of the United States of America* 108, no. 1 (2011): 4680–87.

16. Yatsunenko et al., "Human Gut Microbiome."
17. See Figure 2 in De Wolfe et al., "Chasing Ghosts," 4.
18. Margaret Lock, "The Tempering of Medical Anthropology: Troubling Natural Categories," *Medical Anthropology Quarterly* 15, no. 4 (2001): 478–92.
19. Lock and Nguyen, *An Anthropology of Biomedicine.*
20. Amade M'charek, *The Human Genome Diversity Project: An Ethnography of Scientific Practice* (Cambridge: Cambridge University Press, 2005).
21. https://grants.nih.gov/policy/inclusion/women-and-minorities/guidelines.htm.
22. Michael Yudell, Dorothy Roberts, Rob DeSalle, and Sarah Tishkoff, "Taking Race Out of Human Genetics," *Science* 351, no. 6273 (2016): 564–65.
23. Vann R. Newkirk, "Precision Medicine's Post-racial Promise," *The Atlantic,* June 8, 2016, https://www.theatlantic.com/politics/archive/2016/06/precision-medicine-race-future/486143/.
24. Vence Bonham, Shawneequa L. Callier, and Charmaine D. Royal, "Will Precision Medicine Move Us beyond Race?," *New England Journal of Medicine* 374, no. 21 (2016): 2003–5; Jonathan Kahn, "Revisiting Racial Patents in an Era of Precision Medicine," *Case Western Reserve Law Review* 67, no. 4 (2017): 1153–69.
25. Helmreich, *Sounding the Limits of Life,* 66.
26. Alicia J. Foxx, Karla P. Franco Meléndez, Janani Hariharan, Ariangela J. Kozik, Cassandra J. Wattenburger, Filipa Godoy-Vitorino, and Adam R. Rivers, "Advancing Equity and Inclusion in Microbiome Research and Training," *mSystems* 6, no. 5 (2021).
27. Carlotta De Filippo, Duccio Cavalieri, Monica Di Paola, Matteo Ramazzotti, Jean Baptiste Poullet, Sebastien Massart, Silvia Collini, Giuseppe Pieraccini, and Paolo Lionetti, "Impact of Diet in Shaping Gut Microbiota Revealed by a Comparative Study in Children from Europe and Rural Africa," *Proceedings of the National Academy of Sciences of the United States of America* 107, no. 33 (2010): 14691–96; Andres Gomez, Klara J. Petrzelkova, Michael B. Burns, Carl J. Yeoman, Katherine R. Amato, Klara Vlckova, and David Modry, "Gut Microbiome of Coexisting BaAka Pygmies and Bantu Reflects Gradients of Traditional Subsistence Patterns," *Cell Reports* 14, no. 9 (2016): 2142–53; Simone Rampelli, Stephanie L. Schnorr, Clarissa Consolandi, Silvia Turroni, Marco Severgnini, Clelia Peano, Patrizia Brigidi, Alyssa N. Crittenden, Amanda G. Henry, and Marco Candela, "Metagenome Sequencing of the Hadza Hunter-Gatherer Gut Microbiota," *Current Biology* 25, no. 13 (2015): 1682–93.
28. Erica D. Sonnenburg and Justin L. Sonnenburg, "The Ancestral and Industrialized Gut Microbiota and Implications for Human Health," *Nature Reviews Microbiology* 17, no. 6 (2019): 383.
29. Similarly, some scholars, such as geographer Jamie Lormier, take on the concept of "rewilding" as it relates to a redefinition of nature, and the implications for conservation. Lorimer, *Wildlife in the Anthropocene:*

Conservation after Nature (Minneapolis: University of Minnesota Press, 2015); Lorimer, *The Probiotic Planet.*

30. Nicolai Karcher, Edoardo Pasolli, Francesco Asnicar, Kun D. Huang, Adrian Tett, Serena Manara, Federica Armanini et al., "Analysis of 1321 Eubacterium Rectale Genomes from Metagenomes Uncovers Complex Phylogeographic Population Structure and Subspecies Functional Adaptations," *Genome Biology* 21, no. 138 (2020): 1–27.
31. Karcher et al., 14.
32. M'charek, *Human Genome Diversity Project*; Jenny Reardon, *Race to the Finish: Identity and Governance in an Age of Genomics* (Princeton, N.J.: Princeton University Press, 2005).
33. Cori Hayden, *When Nature Goes Public* (Princeton, N.J.: Princeton University Press, 2003).
34. The publication of the American Association for the Advancement of Science.
35. Jose C. Clemente, Erica C. Pehrsson, Martin J. Blaser, Kuldip Sandhu, Zhan Gao, Bin Wang, Magda Magris et al., "The Microbiome of Uncontacted Amerindians," *Scientific Advances* 3, no. 1 (2015): e1500183.
36. Stephanie Maroney, "Reviving Colonial Science in Ancestral Microbiome Research," *Microbiosocial* (blog), January 10, 2017, https://microbiosocial.wordpress.com/2017/01/10/reviving-colonial-science-in-ancestral-microbiome-research/.
37. David Wendler, E. J. Emanuel, and R. K. Lie, "The Standard of Care Debate: Can Research in Developing Countries Be Both Ethical and Responsive to Those Countries' Health Needs?," *American Journal of Public Health* 94, no. 6 (2004): 923–28.
38. Krithivasan Sankaranarayanan, Andrew T. Ozga, Christina Warinner, Raul Y. Tito, Alexandra J. Obregon-Tito, Jiawu Xu, Patrick M. Gaffney et al., "Gut Microbiome Diversity among Cheyenne and Arapaho Individuals from Western Oklahoma," *Current Biology* 25, no. 24 (2015): 3161–69.
39. Ravel et al., "Vaginal Microbiome of Reproductive-Age Women."
40. Mansfield and Guthman, "Epigenetic Life."
41. Mansfield and Guthman, 6.
42. Ravel et al., "Vaginal Microbiome of Reproductive-Age Women," 4684.
43. Richard A. Cone, "Vaginal Microbiota and Sexually Transmitted Infections That May Influence Transmission of Cell-Associated HIV," *Journal of Infectious Diseases* 210, no. 3 (2014): 616–21.
44. Hanneke Borgdorff, Charlotte van der Veer, Robin van Houdt, Catharina J. Alberts, Henry J. de Vries, Sylvia M. Bruisten, Marieke B. Snijder, Maria Prins, Suzanne E. Geerlings, Maarten F. Schim van der Loeff, and Janneke H. H. M. van de Wijgert, "The Association between Ethnicity and Vaginal Microbiota Composition in Amsterdam, the Netherlands," *PLOS One* 12, no. 7 (2017): 181135; Jennifer M. Fettweis, J. Paul Brooks, Myrna G.

Serrano, Nihar U. Sheth, Philippe H. Girerd, David J. Edwards, Jerome F. Strauss, The Vaginal Microbiome Consortium, Kimberly K. Jefferson, and Gregory A. Buck, "Differences in Vaginal Microbiome in African American Women versus Women of European Ancestry," *Microbiology* 160, no. 10 (2014): 2272–82; Sujatha Srinivasan, Noah G. Hoffman, Martin T. Morgan, Frederick A. Matsen, Tina L. Fiedler, Robert W. Hall, Frederick J. Ross et al., "Bacterial Communities in Women with Bacterial Vaginosis: High Resolution Phylogenetic Analyses Reveal Relationships of Microbiota to Clinical Criteria," *PLoS One* 7, no. 6 (2012): 0037818.

45. Jennifer Fettweis, Myrna G. Serrano, J. Paul Brooks, David J. Edwards, Philippe H. Girerd, Hardik I. Parikh, Bernice Huang et al., "The Vaginal Microbiome and Preterm Birth," *Nature Medicine* 25 (June 2019): 1012–21.
46. Jacques Ravel and Rebecca M. Brotman, "Translating the Vaginal Microbiome: Gaps and Challenges," *Genome Medicine* 8, no. 35 (2016): 1–3.
47. Troy Duster, "The Molecular Reinscription of Race: Unanticipated Issues in Biotechnology and Forensic Science," *Patterns of Prejudice* 40 (2006): 427–41.
48. Duana Fullwiley, "The Molecularization of Race: Institutionalizing Human Difference in Pharmacogenetics Practice," *Science as Culture* 1, no. 16 (2007): 1–30.
49. Paula Braveman, Katherine Heck, Susan Egerter, Tyan Parker Dominguez, Christine Rinki, Kristen S. Marchi, and Michael Curtis, "Worry about Racial Discrimination: A Missing Piece of the Puzzle of Black–White Disparities in Preterm Birth?," *PLoS One* 12, no. 10 (2017): 0186151; Amelia R. Gavin, N. Grote, K. O. Conner, and T. Fentress, "Racial Discrimination and Preterm Birth among African American Women: The Important Role of Posttraumatic Stress Disorder," *Journal of Health Disparities Research and Practice* 11, no. 4 (2018): 91–109; Michael R. Kramer and Carol R. Hogue, "What Causes Racial Disparities in Very Preterm Birth? A Biosocial Perspective," *Epidemiology Review* 31, no. 1 (2009): 84–98.
50. See chapter 3 for a definition.
51. Yatsunenko et al., "Human Gut Microbiome," 222.
52. Ruth Ley and collaborators studied the fecal microbiota of humans and other animals and found that host diet influences bacterial diversity—herbivores had the most diverse populations, followed by omnivores, with carnivore guts being the least microbially diverse. Ruth Ley, M. Hamady, C. Lozupone, P. J. Turnbaugh, R. R. Ramey, J. S. Bircher, M. L. Schlegel et al., "Evolution of Mammals and Their Gut Microbes," *Science* 320, no. 5883 (2008): 1647–51.
53. Fortenberry, "Uses of Race and Ethnicity"; Steve Epstein, *Inclusion: The Politics of Difference in Medical Research* (Chicago: University of Chicago Press, 2007).
54. Helmreich, *Sounding the Limits of Life*, 67.

55. Lock and Nguyen, *An Anthropology of Biomedicine*, 3.
56. Subramaniam and Willey, "Introduction to Science out of Feminist Theory Part One," 12.
57. Barad, *Meeting the Universe Halfway*.
58. Farid Dahdouh-Guebas, J. Ahimbisibwe, Rita Van Moll, and Nico Koedam, "Neo-colonial Science by the Most Industrialised upon the Least Developed Countries in Peer-Reviewed Publishing," *Scientometrics* 56 (2003): 329–43; Foxx et al., "Advancing Equity and Inclusion."
59. Hannah Brown and Alex Nading, "Introduction: Human Animal Health in Medical Anthropology," *Medical Anthropology Quarterly* 33, no. 1 (2019): 5–23.
60. Anthropologist John Hartigan makes a powerful point in his book *Care of the Species* that "race is not uniquely about people." His conclusions push us to think about how racial thinking applied to nonhumans is crucial to understanding race. Hartigan, *Care of the Species* (Minneapolis: University of Minnesota Press, 2017).
61. Yates-Doerr, "Whose Global, Which Health?"; Roberts and Sanz, "Bioethnography."
62. Roberts, "Bioethnography and the Birth Cohort."
63. Biehl and Petryna, *When People Come First*, 4.
64. Fortenberry, "Uses of Race and Ethnicity," 165.
65. This is Karen Barad's invention—an ethics inseparable from being and knowing, especially between scientific practice and the world itself. Barad, *Meeting the Universe Halfway*.
66. https://www.blackinmicrobiology.org/.
67. Shameka P. Thomas, Kiana Amini, K. Jameson Floyd, Rachele Willard, Faeben Wossenseged, Madison Keller, Jamil B. Scott, Khadijah E. Abdallah, Ashley Buscetta, and Vence L. Bonham, "Cultivating Diversity as an Ethos with an Anti-racism Approach in the Scientific Enterprise," *Human Genetics and Genomics Advances* 2, no. 4 (2021): 100052.
68. Thomas et al., 3.
69. Ishaq et al., "Introducing the Microbes and Social Equity Working Group," 472.
70. https://sueishaqlab.org/tag/speaker-series/.
71. Foxx et al., "Advancing Equity and Inclusion in Microbiome Research and Training."
72. Foxx et al., 6.
73. De Wolfe et al., "Chasing Ghosts."
74. Benezra, "Race in the Microbiome."
75. De Wolfe et al., "Chasing Ghosts," 2.
76. De Wolfe et al., 5.

Conclusion

1. Eve Tuck, "Suspending Damage: A Letter to Communities," *Harvard*

Education Review 79, no. 3 (2009): 409–27.

2. Leighton and Roberts, "Trust/Distrust in Multidisciplinary Collaboration."
3. Farmer, "An Anthropology of Structural Violence," 308.
4. Farmer, 307–8.
5. Callard and Fitzgerald, *Rethinking Interdisciplinarity*, 109.

Index

AMBER BENEZRA is assistant professor of science and technology studies at Stevens Institute of Technology.

CPSIA information can be obtained
at www.ICGtesting.com
Printed in the USA
LVHW020913300423
745543LV00002B/8

9 781517 901295